BEING A
SUSTAINABLE FIRM

BEING A
SUSTAINABLE FIRM

TAKEAWAYS FOR A SUSTAINABILITY-ORIENTED MANAGEMENT

MARIA CRISTINA LONGO

Department of Economics and Business, University of Catania, Italy
Department of Educational Science, University of Catania, Italy

ELEONORA CARDILLO

Department of Economics and Business, University of Catania, Italy

ACADEMIC PRESS

An imprint of Elsevier

ELSEVIER

ISBN: 978-0-443-14062-4

For information on all Academic Press publications visit our website at https://www.elsevier.com/books-and-journals

Publisher: Mica Haley
Acquisitions Editor: Kathryn Eryilmaz
Editorial Project Manager: Sara Valentino
Production Project Manager: Gomathi Sugumar
Cover Designer: Matthew Limbert

Typeset by TNQ Technologies

Working together
to grow libraries in
developing countries

www.elsevier.com • www.bookaid.org

Declaration

When rereading the text, the authors utilized ChatGPT tools solely to improve the readability and language of the work. The authors have carefully supervised and checked every single suggestion or revision proposed by ChatGPT. All work has been carefully reviewed and edited by the authors, who are solely responsible for the content.

Contents

Introduction

Being a Sustainable Firm: Takeaways for a Sustainability-Oriented Management addresses the key strategic issues that firms encounter when entering the complex world of sustainability. The sustainability topic is at the center of a broad debate which revolves around the planet protection, respect for human being and environment balance, and responsible use of resources to allow future generations to continue along the valorization of cultural and historical heritage.

The 2030 Agenda for Sustainable Development is an action program for people, planet, and prosperity. Signed in 2015 by the governments of 193 UN member countries, it defines 17 Sustainable Development Goals divided into subobjectives and five areas of intervention (People, Planet, Prosperity, Peace, Partnership). The definition of common goals on relevant development issues—the fight against poverty, the hunger elimination and the climate change challenge—underlines the commitment of all countries and individuals in leading the planet on the sustainability path. The goals achievement underlines the cascade involvement of governments, institutions, local authorities, businesses, communities, and citizens, engaged in the program implementation according to the various levels of responsibility and different degrees of involvement.

The delineation of five areas identifies the action fields. The first area places *People* at the center in fighting poverty and social exclusion with the aim of promoting well-being, health, and dignified conditions for human capital development. The second area focuses on the *Planet* in fighting the loss of biodiversity with the aim of protecting environmental assets and living species through sustainable management of natural resources. The third area concerns *Prosperity* in combating waste and pollution with the aim of establishing sustainable models of production and consumption, guaranteeing employment and organizational well-being. The fourth area considers *Peace* in countering conflict between countries and peoples and different forms of illegality with the aim of promoting an inclusive, fair, and nonviolent society. Lastly the fifth area is dedicated to *Partnership* in countering partisan interests with the aim of taking coordinated and integrated actions to achieve the aforementioned areas of intervention.

The challenges posed by sustainability intrigue management and accountability scholars interested in the impact of sustainability on the firm competitiveness and the methods of reporting economic, social and environmental issues. The multiplicity of objectives, stakeholders, approaches, interests, perspectives, programs and actions, indicators, reporting models, regulations and guidelines makes sustainability a complex area to manage and monitor.

This book, structured into seven chapters, is positioned within this framework to understand the firms' approaches to being sustainable. It provides an organic overview within which to place firms' decisions and actions on the environmental, social, and governance (ESG) dimension. Based on the different models of circular economy, the volume critically analyzes the way in which firms approach sustainability starting from the EU SDG goals contents and sets of indicators for sustainable business (Chapter 1). This book pays particular attention to *SDG 9—Industry, innovation and infrastructure, SDG 11—Sustainable cities and communities*, and *SDG 12—Ensuring sustainable consumption and production patterns*. These three SDGs are among the Green Deal priorities and represent key aspects for business and local territory, giving a different perspectives into the way of pursuing sustainability in production, services, and cities of the future (Chapter 2). Furthermore, the volume unfolds across the broad spectrum of indicators for sustainability assessment (Chapter 3) and delves into international standards and sustainability reporting guidelines, as relevant socio-environmental reporting systems recognized at an international level. Understanding the logic of sustainability reporting and applying sustainable reporting models to specific business areas offers critical insights and application tools for organizations committed to integrating sustainability into their business and creating new sources of value starting from a common vision of sustainable development (Chapter 4). This book highlights these aspects by linking them to the business challenges and sustainability models in sectors particularly interested in sustainable development, including fashion (Chapter 5), tourism (Chapter 6), and public—private partnerships for sustainable local communities (Chapter 7).

This book is a useful tool for students and scholars of managerial disciplines, interested in the topics of innovation management, sustainability-based strategies, sustainable entrepreneurship, socio-environmental reporting systems, and performance evaluation. The delineation of the regulatory framework and sustainability reporting standards within which strategic decisions are placed constitutes a valuable guide for consultants and practitioners interested in deepening corporate sustainability management tools. Additionally, it provides takeaways for managers on sustainable practices implementation and reporting.

CHAPTER ONE

Sustainability: Meaning and boundaries

1. Sustainability: Key concepts and challenges

Sustainability is a constantly changing concept represented by different and mutable meanings and applied in different fields. Several issues emerge in defining sustainability to understand its meaning, the socio-environmental and political reasons from which it originates, the actions to implement it, and the measures to evaluate it (Pezzey, 1989).

The first issues related to the mankind—environment relationship emerged in the early seventies on topics concerning: scarcity and waste of natural resources, waste disposal and recycling, toxicity of industrial products and environmental emergency, and global overpopulation.

The issue of sustainability became central in the 1980s with the publication of the report *Our Common Future (1987)*, drawn up by the World Commission on Environment and Development (WCED), within the framework of the United Nations Environment Program. The document, known as the "Brundtland Report," provides a precise definition of sustainable development where the temporal element takes on a significant meaning for the protection of present and future generations. Specifically, the Commission defines "sustainable" as development that guarantees the needs of current generations without compromising the ability of future generations to meet their own. This definition proposes a long-term environmental strategy reflecting the need to safeguard the planet, improve the quality of life, and guarantee balanced economic growth, favoring more equitable access to resources, full employment, and social progress (Clayton & Radcliffe, 1996; Crowther et al., 2016). Sustainable development is a central theme of the legislation and policies of the European Union, which has set the 2015 International Development Goals (MDGs) through the "Millennium Development Goal," the subsequent 2016 Sustainable Development Goals (SDGs), up to the 2030 Agenda for Sustainable Development.

The guidelines proposed in the report set several goals for achieving sustainability. These include: promoting more balanced development among

Being a Sustainable Firm
ISBN: 978-0-443-14062-4
https://doi.org/10.1016/B978-0-443-14062-4.00003-9

nations through a reduction of poverty and negative pressures on the environment; extending the concept of growth to include not only economic aspects but also notions of equity and ethical-social values; changing consumption patterns to improve the food satisfaction, building, and energy needs; addressing the problem of population growth by improving resource management; developing environmental risk-management technologies and integrating economic, social, and ecological factors into corporate decision-making processes.

To pursue the aforementioned objectives, some essential conditions for sustainable development achievement are considered, such as:
- political responsibility in decision-making processes;
- economic and productive system based on circular economy models;
- social systems capable of maintaining cohesion through a redistribution of the costs and benefits caused by an unequal development;
- technological innovations that allow for an improvement in energy solutions and efficient resource allocation;
- flexible accountability models and self-correcting governments.

In the concept of sustainable development, the intergenerational problem emerges, which implies taking into account the coexistence of different parts: past, present, and future generations. For example, current generations adopt measures to safeguard and protect future generations, a choice that could create a conflicting situation with respect to a sacrifice that subjects face to improve their existence and future living conditions. Furthermore, the generation that today manages socio-environmental questions often assumes responsibility for damages that derive from previous generations. The environmental, economic, and social dimensions characterize the areas of sustainability. Specifically, the environmental dimension is concerned with maintaining the quality and reproducibility of environmental resources. The economic dimension interprets sustainability as the ability to generate income and work to support the population by monitoring the relationship between costs and benefits in production processes. The social dimension is careful to guarantee human well-being and to satisfy the community's needs. Social sustainability is possible when the structures and the complex of corporate relationships are able to create livable and sustainable social systems (Mies & Gold, 2021).

The relationship between economy, society, and environment is complex and is connected to issues related to ethics, cultural values, and social responsibility. It has led to the affirmation of various alternative approaches, programs, and actions to traditional economic models, including the Green

Economy and the Circular Economy Model. According to what has been outlined by the United Nations Environment Program, the Green Economy is an economy with low carbon emissions, efficient in the use of resources, and socially inclusive. It argues that the ecological processes of natural and seminatural systems can be exploited for the benefit of human beings without endangering the sustainability of ecosystems (Van Loon et al., 2015; D'Amato & Korhonen, 2021). The circular economy is an evolution of the green economy (Adams, 2019), which is based on an economic model capable of allowing for smart, inclusive, and sustainable growth (Fay, 2012; Kovacic et al., 2019). The term circular economy indicates a self-regenerating economy in which innovative management of waste, through the reuse of materials in subsequent production cycles and the waste reduction in the perspective of zero waste, is the basis of widespread and lasting sustainability (Braungart & McDonough, 2002; Pearce & Turner, 1990). This sustainability can be achieved through the efficient use of resources, the reuse of production factors, and the use of renewable energy sources, which is summarized in the pursuit of the threeRs namely Reduce, Reuse, and Recycle (Ioannidis et al., 2021; Kabirifar et al., 2020). Circular economy indicators pursue concrete actions and measurable results. The measure of the circularity of economic activities evaluates sustainable performance through standardized balance sheets and reports which constitute an essential basis for setting priorities, objectives, and monitoring actions.

The European Commission is careful to monitor the progress and critical issues of the transition toward a circular economy. In 2018 the Communication on a monitoring framework for the circular economy from the Commission to the European Parliament, the Council, the European Economic and Social Committee, and the Committee of the Regions was presented. It represents the official tool used to monitor circularity through key indicators and is the most representative of the circular economy. Fig. 1.1 shows the circular economy monitoring framework. The indicators are divided into four categories and refer to the entire life process of resources, products, and services: production and consumption, waste management, secondary raw materials, and competitiveness and innovation.

Eurostat has created a further dataset of sustainable development and the circular economy, such as the Resource Efficiency Scoreboard, Raw Materials Scoreboard, and Environmental accounts. The first proposes a roadmap for an efficient Europe COM (2011/571) and for the Europe 2020 Strategy. Raw Materials Scoreboard offers a vision of the challenges and opportunities relating to the entire raw materials production chain with the aim of

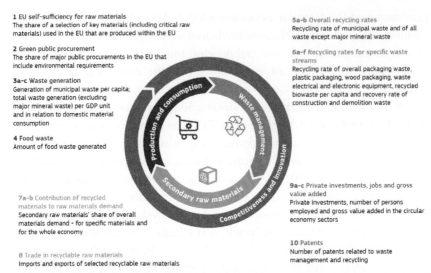

1 EU self-sufficiency for raw materials
The share of a selection of key materials (including critical raw materials) used in the EU that are produced within the EU

2 Green public procurement
The share of major public procurements in the EU that include environmental requirements

3a–c Waste generation
Generation of municipal waste per capita; total waste generation (excluding major mineral waste) per GDP unit and in relation to domestic material consumption

4 Food waste
Amount of food waste generated

7a–b Contribution of recycled materials to raw materials demand
Secondary raw materials' share of overall materials demand – for specific materials and for the whole economy

8 Trade in recyclable raw materials
Imports and exports of selected recyclable raw materials

5a–b Overall recycling rates
Recycling rate of municipal waste and of all waste except major mineral waste

6a–f Recycling rates for specific waste streams
Recycling rate of overall packaging waste, plastic packaging, wood packaging, waste electrical and electronic equipment, recycled biowaste per capita and recovery rate of construction and demolition waste

9a–c Private investments, jobs and gross value added
Private investments, number of persons employed and gross value added in the circular economy sectors

10 Patents
Number of patents related to waste management and recycling

Figure 1.1 Circular economy monitoring framework. *Source: From Communication of the regions on a monitoring framework for the circular economy, 2018.*

supporting policies and choices of the various economic sectors (www.ec. europa.eu). Environmental accounts detect information on the environment and on the impact of the economy on the environment through the use of physical units to record flows of materials and energy juxtaposed with economic values.

The Global Resources Outlook published by the UN in 2020 shows that in 50 years the flows of matter (biomass, metals, nonmetallic minerals, and fossil fuels extracted from the earth) have tripled leading to unsustainable growth. Indeed, global extraction of raw materials has increased 3.4 times since 1970, from 27 to 92 billion tons per year, compared with the increase in global population, which has doubled. The growth in natural resource consumption was absorbed by upper-middle-income countries, which achieved a 56% global share of material consumption in 2017. On a per-capita basis, material consumption levels in high-income countries are 60% higher than those in upper-middle-income countries and 13 times the level of low-income countries (United Nations Environment Programme UNEP - International Resources Panel)

Ultimately, the data relating to the flows of materials for industrial processing at a global and European level highlight a rather limited transition process to circular economy models. Currently, circularity is more developed at the company level or within specific business sectors or production chains. The challenges for the transition to sustainability require strategies and policies promoted at the European and national levels to accelerate the diffusion of sustainable business models in companies and organizations to improve the efficiency and competitiveness of the markets.

The Blue Economy introduced by the Belgian economist Gunter Pauli (2010) addresses the issue of water sustainability in harmony with the principles of the green economy. Unlike this which aims to stimulate investments in technologies to reduce the environmental impact, the blue economy pursues the goal of completely eliminating emissions. This emerging discipline, based on biomimicry, is concerned with analyzing and reproducing the biological and biomechanical processes of terrestrial flora and fauna. It represents an innovative approach to the management of seas and oceans and envisages a series of policies aimed at supporting entrepreneurial activities that simultaneously pursue economic, social, and environmental objectives. In particular, the blue economy offers advanced techniques for the production and transformation of materials with zero emissions through the observation of how nature works. The Eastgate Building Center in Zimbabwe is an example of a blue economy building, designed by architect Mick Pearce according to specific mathematical models that exploit the construction principles of termite mounds, capable of self-cooling without any ventilation system.

2. The main sustainability regulations

The 2030 Agenda sets 17 sustainable development goals that provide a shared global project, applicable on the basis of the country's characteristics, where the three dimensions of sustainability linked to economic, social, and environmental development are integrated (Cardillo & Longo, 2020).

Reconstructing the regulatory path, the first official recognition of the concept of sustainable development occurred in the 1992 Earth Summit in Rio de Janeiro where it was established that environmental issues and possible solutions should involve all the countries of the world. The UN World Commission on the Environment and Development, addressing global economics, environment, and development, proposed a sustainable economy model aimed at satisfying the needs of current generations without compromising the planet's resources' ability to meet those of future generations. Several action programs have been issued, including: Agenda 21; the declaration of principles for sustainable forest management; the framework convention on climate change; the Framework Convention on Biodiversity; and the Rio Declaration on Environment and Development. Of note is the "United Nations Framework Convention on Climate Change" held in New York on 9 May 1992 on the subject of climate change and polluting emissions. The art. 2 of the Convention on Climate Change establishes that the primary objective is "to stabilize, in accordance with the relevant provisions of the Convention, the concentrations of greenhouse gases in the atmosphere at a level such that any dangerous interference by human activities on the climate system is excluded. This level should be achieved within a sufficient timeframe to allow ecosystems to adapt naturally to climate changes and to ensure that food production is not threatened, and economic development can continue at a sustainable peace." The subsequent Kyoto Protocol (1997) is an international agreement signed by more than 160 countries concerning actions to be taken to limit climate change. In this agreement, the industrialized countries undertook to reduce their total greenhouse gas emissions by at least 5% in the period 2008—12.

In 2002, the UN summit held in Johannesburg on sustainable development undertook concrete actions focusing on the relationship between economic development, environmental sustainability, and social development attributable to the concept of the Triple Bottom Line (Bennet & James, 1998; Elkington, 1994, 1997, 1999). In its general sense, the Triple Bottom Line (TBL) jointly considers economic value with environmental and social aspects, anchoring it to the dimensions of sustainability. The path leading to TBL brings about a change in the production methods of companies by directing their processes toward respect for socio-environmental variables and toward partnership. The sustainable agenda has also changed the way of conceiving and managing the temporal logic of action, being more oriented toward results and long-term projects; this logic leads businesses and public administrations to formulate choices that concern the future and the growth of future generations. The approach stimulates a rethinking of corporate governance logics and structures to guarantee the pursuit of an adequate level of triple bottom line. Economic, social, and environmental development becomes interdependent and mutually reinforcing components for sustainability. The economic aspect contributes through the management and allocation of resources to the growth of society; the social aspect highlights the quality of life of the subjects by assessing equity between peoples, communities, and nations at a macro level; the environmental aspect aims at strengthening the activities of protection and conservation of the natural environment.

In 2006, the Sustainable Development Strategy (Council of the European Union), implementing the Sustainable Development Strategy (SSS) adopted by the European Council in Gothenburg in 2001 and the Johannesburg action program in 2002, strengthened the objectives of environmental, economic, and social sustainability by focusing on investments in R&D, rural development programs and agriculture, on the role of science and innovation, climate security, and reduction of dependence on fossil fuels. In 2012, at the Doha UN conference on climate change, the agreements of the Kyoto Protocol were reworked, providing for more binding actions and practices for the reduction of emissions and the promotion of the green economy extended until 2020.

Also in 2012, the UN Conference on Environment and Development in Rio de Janeiro dealt with elaborating a new global strategy focusing essentially on two relevant areas such as the Green Economy in the context of sustainable development and the fight against poverty and the formulation of a framework institution to ensure environmental, social, and economic sustainability.

In 2015, the European Commission adopted a "Circular Economy Package" which will enter into force in 2030. The package focused on the life cycle of products (collection, recycling, and recovery of materials) and the efficient use of resources in the typical phases of production (procurement, design, production, distribution, use, and disposal) with the aim of increasing the recycling rate in the ember countries to reach at least 70% of municipal waste recycling and 80% of packaging waste with prohibition to send biodegradable and recyclable products to landfills. In March 2020, the European Commission proposed "A new action plan for the circular economy" with the aim of enhancing sustainable products and making users responsible for the green economy. In February 2021, the European Commission envisaged more incisive rules on the recycling of products and on the use and consumption of materials. In 2022 it proposed a revision of the rules on persistent organic pollutants for the reduction of chemicals in waste and production and new rules on packaging to improve its design, traceability, reuse, and recycling. The 2030 Agenda has been the latest action program signed in 2015 by the governments of 193 UN member countries. It established 17 objectives for sustainable development (Sustainable Development Goals, SDGs), which provided for a structured and complex action program of 169 targets, starting in 2016 and a commitment to achieve them by 2030. The sustainability agenda has led collaboration between governments and organizations within society to implement strategies on key sustainability issues.

In 2019, the European Commission drafted the European Green Deal, as an integral part of the implementation strategy of the 2030 Agenda. The primary objective was to reshape the EU into a fair, prosperous, and competitive society, with zero greenhouse gas emissions. Apart from addressing environmental and climate goals, the Green Deal advocated for a shift towards sustainability, emphasizing equity, justice, and social participation. It positioned the well-being of citizens at the core of European economic policy. The implementation of these programs, coinciding with the outbreak of the pandemic, represents a driving force to promote economic recovery toward a more resilient, sustainable, and equitable Europe. The European Green Deal played a crucial role as an inclusive and sustainable growth strategy, grounded in the green transition, digital transformation, circular economy, and common agricultural and social cohesion policies. The recovery and reconstruction package requested by the European Parliament in its resolution of April 17, 2020, positioned the European Green Deal at the heart of relaunching the economy, preserving strategic EU industrial sectors, generate employment, and accelerating the ecological transition. These aims were further reinforced by the Parliamentary Resolution of May 15, 2020, which underlines the need to give rapid and effective responses to the social emergencies arising from the pandemic and lockdown, fostering transformative resilience capacity. To this end, the European Commission proposed on May 27, 2020, the "Next Generation EU," representing the European emergency instrument for recovery with the allocation of EUR 750 billion for the member countries. Funding is contingent on the development of national recovery plans aligned with national energy and climate plans, just transition plans, partnership agreements, and operational programs under EU funds. Public measures to revitalize the economy are planned on the basis of economic multipliers of spending in the short and medium–long term, with the additional benefit of contributing to the fight against climate change. Industrial strategies and investment plans geared towards reducing financial risks will have to comply with minimum environmental and social standards to achieve climate and environmental objectives in accordance with the European taxonomy of sustainable finance (Fig. 1.2).

Fig. 1.2 summarizes the key points of the Sustainable Europe Investment Plan, which include: toxic-free environment; protection and restoration of ecosystems and biodiversity; fair, healthy, and ecological food system; sustainable and intelligent mobility; energy efficiency; industry for a clean and circular economy; clean, cheap, and secure energy; More ambitious EU climate targets for 2030 and 2050. The Sustainable Europe Investment

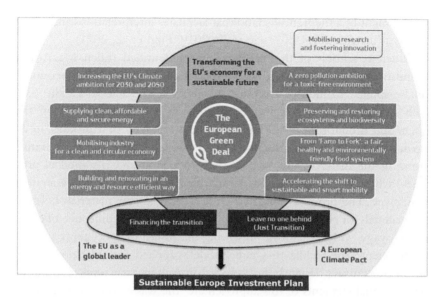

Figure 1.2 European Green Deal Investment Plan. *Source: From Communication from the commission to the European Parliament, the Council, the European Economic and Social Committee and the Committee of the Regions, 2020: 2.*

Plan will stimulate private and public sustainable investments in the climate, environment, and society sectors.

3. Circular economy models and approaches

The circular economy, as a restorative and regenerative economy, re-places the concept of product end-of-life with the recovery and reuse of the same through the design of materials, components, products, systems, and business models based on the use of renewable energy and on the elimination of toxic chemicals and waste (MacArthur, 2013; Stahel, 2019). The circular model goes beyond the traditional linear economic model in which the life cycle of a product is characterized by the phase of extraction, production, consumption, and disposal. In the circular economy, the problems related to environmental unsustainability and pollution and waste production are addressed through a circular model that exploits the interconnections between design, production, consumption, and disposal by involving the various players in the production system (Geissdoerfer et al., 2017).

The need to bring policies, guidelines, and action programs based on circular economy models back to a common framework is leading to the development of a theoretical strand that facilitates interpretations and applications at the firm level (Kovacic et al., 2019; Pizzi et al., 2021).

Different schools of thought have been identified (Ellen MacArthur Foundation, 2020) to explain how the principles of the circular economy can be applied to industrial production processes. Cradle to Cradle (i.e., "C2C," "regenerative design") is an innovative approach based on the design and monitoring of the entire production chain of a product in an eco-sustainable key. The principle behind this approach is the regenerative nature of the materials involved in the production and commercial processes, which are assimilated to natural elements capable of regenerating themselves. The products obtained with eco-sustainable materials become raw materials for new products. The approach pursues the objective of eliminating waste with the aim of preserving the health of ecosystems and the environment. The Cradle-to-Cradle model is applied not only to industrial and manufacturing design but also to urban centers, structures, and buildings. Essential elements of this approach are: the waste cycle, in which the components of a product can be reused, becoming nutrients for another production chain; the energy used is of the renewable type and replaces the fossil-origin energy; diversity, which underlines the commitment of companies in the field of equality, social parity, and biodiversity. This approach involves a production model that takes into account not only form and functionality but also quality and safety in the various cycles of use. The trademark registered by McDonough Braungart Design Chemistry is used by companies to obtain Cradle-to-Cradle certification, for improving their brand reputation. The distinctive features of Cradle-to-Cradle certified products are high economic efficiency, low environmental pollution, the particularity of the design, the composition of the product, functionality, quality, and safety in the reuse cycles.

The *Performance Economy*, introduced by Walter Stahel (Stahel, 2019) represents a form of circular economy based on designing the production process in a closed loop. The performance is developed in the following areas: *Producing Performance*, which analyzes the measurement of the creation of wealth with respect to the resources consumed; *Selling Performance*, which identifies the business models capable of exploiting the synergies derived from increased work compared to the reduced resource consumptions; *Managing Performance*, over time to measure the relationship between job creation and resource consumption. The *Performance Economy* therefore pays attention to the outcomes of the production process, considering the extension of the product life cycle, the durability of goods, and the renewal and waste reduction. *Biomimicry*, literally imitation of life, takes nature as a model for developing ideas, innovations, and processes and evaluating their sustainability based on the ecological standard. Nature is observed in terms of functional efficiency, energy efficiency, self-adaptation, sharing, and recycling. *Industrial Ecology* deals with the study of matter and energy flows in industrial systems based on the recovery of waste for use as input in developing industrial plans. *Natural capitalism* refers to natural assets and is based on the principle of increasing the productivity of natural resources, adopting biological models and materials and reinvesting in natural capital. Finally, *Regenerative Design*, inspired by agriculture, applies the principle of the regenerative nature of products and resources to the various production sectors.

These models are reflected in the circular economy *"Butterfly Diagram"* promoted by the Ellen MacArthur Foundation (2019), which reconstructs the starting point for a regenerative design. This diagram illustrates the continuous flow of technical (left) and biological (right) materials according to the circle of value. The three principles of action of the circular economy aim to keep products, components, and materials at their highest value content in both cycles. The first principle is to conserve and enhance natural resources by controlling the withdrawal of nonrenewable resources and balancing the flow of renewable resources according to different resolution levers, such as regeneration, virtualization, and exchange. The second principle is to optimize resource yields in both by deriving maximum utility from them in the technical and biological cycles using processes of regeneration, sharing, optimization, and execution of the cycles. Finally, the third principle is to foster system effectiveness by systematically identifying and eliminating negative externalities and using all available resolution levers to remove them.

The central part of the diagram shows the two vertical flows of the linear model (extraction, production, consumption, and disposal of materials and the energy used). On the left is the technical cycle within which materials, components, and products are kept in use in the economic system at their state of maximum value for as long as possible. The technical cycles also include products created with nonbiodegradable resources (metal, artificial and synthetic fabrics, plastic). The technical cycle is set up to recover and restore products through strategies such as reuse, repair, remanufacturing, and recycling. On the right is the biological cycle where the nutrients are put back into the biosphere, rebuilding the so-called natural capital through composting and anaerobic digestion, which allows for the extraction of fuel and precious nutrients such as nitrogen, phosphorus, potassium, and micronutrients. Within the biological cycle, there are organic materials (food, natural textile fibers, wood) as they are biodegradable and renewable by nature. The lengthening of their duration takes place through the "cascading" process (creation of waste) where the organic materials are reused in different areas until they are composted and subjected to the anaerobic digestion process when they can no longer be used. Through the circular economy model, firms can minimize waste, raw material costs, and the length of the supply chain.

4. Framing the sustainability strategy: Being born versus becoming a sustainable firm

A firm oriented toward sustainability bases its entire strategy on the concept of regeneration. Its strategic mission is to create economic, environmental, and social value by promoting circular economy models. The entire production life cycle of the sustainable firm's product promotes a system based on the waste management of waste and its reuse until the end of its life. The production cycles are characterized by the reuse of products that return to the customers, reconditioning through which the products are returned to the producer's sphere for new uses, the regeneration through the transition to a new manufacturing process, and recycling through the reuse of the materials throughout the entire process.

Sustainability strategies can represent the basis for a competitive cost and differentiation advantage capable of generating an increase in profitability in the medium–long term, with a positive correlation between sustainability and economic-financial performance (Flammer, 2015; Walsh & Dodds, 2017). The literature has found that sustainability in the ethical, social, and environmental fields allows companies to limit the risks associated with the occurrence of events that can damage the corporate reputation (White, 2009). Other authors (León-Soriano et al., 2010; Teh & Corbitt, 2015) have underlined that sustainability is a distinctive element for market positioning, strengthening its reputation and brand. The orientation towards sustainability allows firms to improve business productivity and corporate environment if based on an improvement in working conditions, the regenerative nature of raw materials, and respect for ethical values.

Firms oriented toward sustainable markets develop a competitive advantage by co-creating sustainable markets or by shifting market segments by intercepting niches of consumers sensitive to ethics and eco-sustainable products (Eide et al., 2020; Saxena & Khandewal, 2010). Sustainability increases a positive attitude toward social change and has a positive impact on customer retention and customer experience (Corsini et al., 2020; Testa et al., 2020; Yazdanifard & Mercy, 2011). The way in which firms commit themselves to implement sustainability actions is essential for answering effectively to market emerging needs (Loorbach and Wijsman, 2013). In terms of the territory, a sustainability-oriented firm creates value for local communities by contributing to the improvement of the quality of life, the protection of the environment, and social well-being.

However, being sustainable is very different from becoming sustainable. A firm that is born sustainable designs its business by evaluating not only the social, environmental, and economic impact of a new product but also by activating a dynamic circuit that gives life to a resilient and sustainable ecosystem, creating long-term value (Dwivedi et al., 2019). The challenge for the new venture is to design products that enhance community well-being by envisioning the business as a "living system" (Bellato et al., 2022). Key elements of the living system include the durability of components, biodegradability, use of modular and decomposable parts, the reuse of raw materials following the principles of production waste management, and the analysis of return flows from returned and end-of-life products (Díaz-García et al., 2015). Furthermore, the sustainability configuration should fit into the management dynamics, refining it over time. To achieve this, strategies and specific areas of action are required, making the choices and activities sustainable. Pursuing a sustainability strategy goes beyond explicit commitments and actions; it also entails taking an active role in promoting sustainability. Specifically, a sustainability-oriented new venture involves adopting sustainability reports and external guarantee procedures (Jones et al., 2014). The implementation of sustainable practices can find barriers that can slow down this process. This is sometimes due to a lack of knowledge of the ways in which it is necessary to focus on sustainability.

Firms already operating in the market are asking themselves the strategic question of introducing new paths and competitive levers in line with sustainable development (Gray & Bebbington, 2000). This requires a change in the objectives set toward sustainable growth, a review of the overall image of the company with an ethical-social imprint, and the adoption of a system thinking approach that adds energy and ecology into corporate decisions and budgets. This leads firms to activate sustainable transition strategies, moving from linear to circular models that promote regenerative cycles with reduced environmental impact. The measurement of a multifaceted performance raises some critical issues due to the fact that, conducting an adequate reporting and evaluation activity that reflects the purposes expressed by the integration of the three dimensions of the Triple Bottom Line has proven to be difficult (Gray & Bebbington, 2000). Measuring the three dimensions in an integrated and coherent manner according to a common unit of measurement and analysis is a complex operation. For instance, when the TBL is set in financial terms, it becomes necessary to quantify different aspects in monetary terms. This criterion poses delicate measurement problems because most of the costs and benefits deriving from the social and environmental dimension are characterized by intangibility. Consequently, converting these aspects into measurable quantities and values becomes a complex task.

Sustainability transitions require a "long-term, multidimensional and fundamental transformation processes through which established sociotechnical systems shift to more sustainable modes of production and consumption" (Markard et al., 2012, p. 956). The literature on sustainability transition strategy (Geels, 2019) highlights that the path toward sustainability aims to strike between economic performance and the need to protect the territory and the surrounding environment. Some studies (Vinnari & Vinnari, 2014) analyze the obstacles when deviating from its various dimensions and propose actions to respond to these obstacles. Others (Schäpke et al., 2017) interpret the transition to sustainability as a profound process of change in social systems where social learning and the development of social capital are essential elements.

Transition processes identify the relationship between sustainability goals and stakeholder involvement over time. Firms moving toward sustainability have to consider how sustainable development initiatives are perceived and how to evaluate the level of awareness and knowledge among the involved actors. The sustainability transition theoretical framework considers the processes of change connected to the dialectic involvement of the interested parties and the implementation of policies (Gössling et al., 2012). Another relevant element is understanding the power relationships between actors in sustainability transitions (Avelino & Wittmayer, 2016). This implies making a distinction between different types and levels of actors, as changes in their power relations affect the understanding of transitional politics. This leads to investigating transition processes with respect to different themes and geographical areas, as sustainability challenges are closely connected to complex factors, such as the socioeconomic context, values and ethics, institutions, companies, and civil society (Geels, 2019). The transition within these relationship systems can present an opportunity to develop new prospects for economic growth based on sustainability. Social and environmental changes are ushering a new phase of corporate responsibility, leading firms to proactively engage with transitions paths (Loorbach & Wisjman, 2013; Witkowska, 2016).

References

Adams, B. (2019). *Green development: Environment and sustainability in a developing world.* Routledge.

Avelino, F., & Wittmayer, J. M. (2016). Shifting power relations in sustainability transitions: A multi-actor perspective. *Journal of Environmental Policy and Planning, 18*(5), 628−649.

Bellato, L., Frantzeskaki, N., Briceño Fiebig, C., Pollock, A., Dens, E., & Reed, B. (2022). Transformative roles in tourism: Adopting living systems' thinking for regenerative futures. *Journal of Tourism Futures, 8*(3), 312−329.

Bennet, M., & James, P. (1998). *The green bottom line: Environmental accounting for management: Current practice and future trends.* Sheffield: Greenleaf Publishing.

Braungart and McDonough. (2002). *Cradle to cradle: Remaking the way we make things.*

Cardillo, E., & Longo, M. C. (2020). Managerial reporting tools for social sustainability: Insights from a local government experience. *Sustainability, 12*(9), 3675.

Clayton, A. M. H., & Radcliffe, J. N. (1996). *Sustainability: A systems approach, London.*

Corsini, F., Gusmerotti, N. M., & Frey, M. (2020). Consumer's circular behaviors in relation to the purchase, extension of life, and end of life management of electrical and electronic products: A review. *Sustainability, 12*(24), 10443.

Council of the European Union. (2006). Renewed EU strategy for Sustainable Development. https://data.consilium.europa.eu/doc/document/ST-10917-2006-INIT/en/pdf. (Accessed 30 April 2023).

Crowther, S., Hunter, B., McAra-Couper, J., Warren, L., Gilkison, A., Hunter, M., … Kirkham, M. (2016). Sustainability and resilience in midwifery: A discussion paper. *Midwifery, 40*, 40−48.

Díaz-García, C., González-Moreno, Á., & Sáez-Martínez, F. J. (2015). Eco-innovation: Insights from a literature review. *Innovation, 17*(1), 6−23.

D'Amato, D., & Korhonen, J. (2021). Integrating the green economy, circular economy and bioeconomy in a strategic sustainability framework. *Ecological Economics, 188*, 107143.

Dwivedi, A., Agrawal, D., & Madaan, J. (2019). Sustainable manufacturing evaluation model focusing leather industries in India: A TISM approach. *Journal of Science and Technology Policy Management, 10*(2), 319−359.

Eide, A. E., Saether, E. A., & Aspelund, A. (2020). An investigation of leaders' motivation, intellectual leadership, and sustainability strategy in relation to Norwegian manufacturers' performance. *Journal of Cleaner Production, 254*, 120053.

Elkington, J. (1997). *Cannibals with forks: The triple bottom line of 21st century business.* Oxford: Capstone.

Elkington, J. (1994). Towards the sustainable corporation: Win-win-win business strategies for sustainable development. *California Management Review, 36*(2).

Elkington, J. (1999). *Triple bottom line revolution—Reporting for the third Millennium.* Australian CPA.

Ellen MacArthur Foundation. (2019). Circular economy systems diagram. https://www.ellenmacarthurfoundation.org. (Accessed 30 April 2023).

Ellen MacArthur Foundation. (2020). Schools of thought that inspired the circular economy. https://www.ellenmacarthurfoundation.org/schools-of-thought-that-inspired-the-circular-economy. (Accessed 30 April 2023).

Fay, M. (2012). *Inclusive green growth: The pathway to sustainable development.* World Bank Publications.

Flammer, C. (2015). Does corporate social responsibility lead to superior financial performance? A regression discontinuity approach. *Management Science, 61*(11), 2549–2568.

Gössling, S., Hall, C. M., Ekström, F., Engeset, A. B., & Aall, C. (2012). Transition management: A tool for implementing sustainable tourism scenarios? *Journal of Sustainable Tourism, 20*(6), 899–916.

Geels, F. W. (2019). Socio-technical transitions to sustainability: A review of criticisms and elaborations of the multi-level perspective. *Current Opinion in Environmental Sustainability, 39*, 187–201.

Geissdoerfer, M., Savaget, P., Bocken, N. M., & Hultink, E. J. (2017). The circular economy—A new sustainability paradigm? *Journal of Cleaner Production, 143*, 757–768.

Gray, R., & Bebbington, J. (2000). Environmental accounting, managerialism and sustainability: Is the planet safe in the hands of business and accounting?. In *Advances in environmental accounting and management* (Vol. 1, pp. 1–44) Emerald Group Publishing Limited.

Ioannidis, A., Chalvatzis, K. J., Leonidou, L. C., & Feng, Z. (2021). Applying the reduce, reuse, and recycle principle in the hospitality sector: Its antecedents and performance implications. *Business Strategy and the Environment, 30*(7), 3394–3410.

Jones, P., Hillier, D., & Comfort, D. (2014). Sustainability in the global hotel industry. *International Journal of Contemporary Hospitality Management, 26*, 5.

Kabirifar, K., Mojtahedi, M., Wang, C., & Tam, V. W. (2020). Construction and demolition waste management contributing factors coupled with reduce, reuse, and recycle strategies for effective waste management: A review. *Journal of Cleaner Production, 263*, 121265.

Kovacic, Z., Strand, R., & Völker, T. (2019). *The circular economy in Europe: Critical perspectives on policies and imaginaries.* Routledge.

León-Soriano, R., Jesús Muñoz-Torres, M., & Chalmeta-Rosaleñ, R. (2010). Methodology for sustainability strategic planning and management. *Industrial Management and Data Systems, 110*(2), 249–268.

Loorbach, D., & Wijsman, K. (2013). Business transition management: Exploring a new role for business in sustainability transitions. *Journal of Cleaner Production, 45*, 20–28.

MacArthur, E. (2013). Towards the circular economy. *Journal of Industrial Ecology, 2*(1), 23–44.

Markard, J., Raven, R., & Truffer, B. (2012). Sustainability transitions: An emerging field of research and its prospects. *Research Policy, 41*(6), 955–967.

Mies, A., & Gold, S. (2021). Mapping the social dimension of the circular economy. *Journal of Cleaner Production, 321*, Article 128960. https://doi.org/10.1016/j.jclepro.2021.128960. In this issue.

Pauli, G. (2010). *The blue economy: 10 years, 100 innovations, 100 million jobs.* (pp. 1–303) Taos, New Mexico, USA: Paradigm publications.

Pearce, D., & Turner, R. K. (1990). *Economics of natural resources and the environment.* Baltimore: Johns Hopkins University Press.

Pezzey, J. (1989). *Definitions of sustainability.* 9. UK: CEED.

Pizzi, S., Corbo, L., & Caputo, A. (2021). Fintech and SMEs sustainable business models: Reflections and considerations for a circular economy. *Journal of Cleaner Production, 281*, 125217.

Saxena, R., & Khandewal, P. K. (2010). Sustainable development through green marketing: The industry perspective. *International Journal of Environmental, Cultural, Economic and Social Sustainability, 6*(6), 59–79.

Schäpke, N., Omann, I., Wittmayer, J. M., Van Steenbergen, F., & Mock, M. (2017). Linking transitions to sustainability: A study of the societal effects of transition management. *Sustainability, 9*(5), 737.

Stahel, W. R. (2019). *The circular economy: A user's guide.* Routledge.

Teh, D., & Corbitt, B. (2015). Building sustainability strategy in business. *Journal of Business Strategy, 36*(6), 39–46.

Testa, F., Iovino, R., & Iraldo, F. (2020). The circular economy and consumer behaviour: The mediating role of information seeking in buying circular packaging. *Business Strategy and the Environment, 29*(8), 3435–3448.

United Nations Environment Programme (UNEP) - International Resources Panel. (2019). Global Resources Outlook 2019. Natural resources for the future we want. https://www.resourcepanel.org/sites/default/files/documents/document/media/unep_252_global_resource_outlook_2019_web.pdf. (Accessed 30 April 2023).

Van Loon, P., Deketele, L., Dewaele, J., McKinnon, A., & Rutherford, C. (2015). A comparative analysis of carbon emissions from online retailing of fast-moving consumer goods. *Journal of Cleaner Production, 106*, 478–486.

Vinnari, M., & Vinnari, E. (2014). A framework for sustainability transition: The case of plant-based diets. *Journal of Agricultural and Environmental Ethics, 27*, 369–396.

Walsh, P. R., & Dodds, R. (2017). Measuring the choice of environmental sustainability strategies in creating a competitive advantage. *Business Strategy and the Environment, 26*(5), 672–687.

White, P. (2009). Building a sustainability strategy into the business. Corporate Governance. *The International Journal of Business in Society, 9*(4), 386–394.

Witkowska, J. (2016). Corporate social responsibility: Selected theoretical and empirical aspects. *Comparative Economic Research. Central and Eastern Europe, 19*(1), 27–43.

Yazdanifard, R., & Mercy, I. E. (2011). The impact of green marketing on customer satisfaction and environmental safety. *International Conference on Computer Communication and Management, 5*, 637–641.

CHAPTER TWO

EU SDG goals: Content and monitoring for sustainable business

1. EU SDG goals and Green Deal

The Sustainable Development Goals (SDGs) introduced by the United Nations in 2015 represent an epic turning point in the realm of sustainability. They draw up a global roadmap that promotes 17 objectives and 169 related goals in a synergistic and integrated way, with the aim of protecting the planet, reducing poverty, hunger, and social inequalities, addressing discrimination between genders, and tackling climate change. The 2030 Agenda is designed in such a way that actions in different areas mutually influence in the pursuit of social, economic, and environmental sustainability. All countries, regardless of their level of economic development, are called upon to pursue the SDGs, making the financial resources, know-how, technology, specific government policies, and action plans available. Monitoring the SDGs is an essential component of the 2030 Agenda as progress in sustainable development is assessed through a set of official EU indicators. Fig. 2.1 represents the set of the 17 SDGs.

SDG 1—No poverty. SDG 1 calls for the eradication of poverty in all its forms and the satisfaction of basic needs. It aims to achieve shared economic prosperity, improving the living standards and the social level of the poorest and most vulnerable sections of the population.

SDG 2—Zero hunger. SDG 2 aims to eliminate hunger and malnutrition by ensuring access to safe, nutritious, and sufficient food. Achieving this goal requires sustainable production systems and investments aimed at promoting agricultural development and rural infrastructure. SDG 2 Monitoring Indicators include the obesity rate and the environmental impacts of agricultural production on water, land, and air.

SDG 3—Good health and well-being. SDG 3 aims to ensure health and social well-being by improving reproductive health and reducing the spread of communicable, noncommunicable, and mental diseases. It also aims to enhance access to healthcare and reduce health risk factors. SDG 3

Being a Sustainable Firm
ISBN: 978-0-443-14062-4
https://doi.org/10.1016/B978-0-443-14062-4.00005-2

Figure 2.1 The UN sustainable development goals. *Source: From Eurostat (2023, p. 21.).*

Monitoring is based on indicators related to healthy living, health determinants, and causes of death.

SDG 4—Quality education. SDG 4 aims to ensure access to quality education for all ages and equip more people with the skills needed to foster employment and entrepreneurship. Monitoring indicators look at basic education, tertiary education, youth and adult learning, and digital skills.

SDG 5—Gender equality. SDG 5 aims to achieve gender equality by eliminating forms of discrimination, violence, and practices harmful to women and girls. The goal also ensures the full participation of women and equal opportunities at all decision-making levels. Monitoring indicators consider gender-based violence, employment opportunities, equal payment, and balanced representation in leadership positions.

SDG 6—Clean water and sanitation. SDG 6 aims to ensure universal access to safe drinking water, sanitation, and hygiene. It also aims to increase the supply of fresh water, improve its quality and enhance the efficiency of use. Monitoring focuses on sanitation, water quality, and water scarcity.

SDG 7—Affordable and clean energy. SDG 7 intends to ensure universal access to affordable and sustainable energy by improving energy efficiency, increasing the share of renewable energy, and diversifying the energy mix. Monitoring indicators look at energy consumption, energy supply, and affordable prices of energy.

SDG 8—Decent work and economic growth. SDG 8 promotes sustained economic growth and high levels of economic productivity with quality jobs, full employment, and decent work. Indicators monitor trends in sustainable economic growth, the employment rate, and decent work quality.

SDG 9—Industry, innovation, and infrastructure. SDG 9 promotes resilient and sustainable infrastructure, inclusive and green industrialization, and research and innovation to address social, economic, and environmental challenges. Monitoring indicators analyze R&D expenditure and personnel, number of patent applications, air emissions from industry, and sustainable shares of passenger and freight transport.

SDG 10—Reduce inequalities. SDG 10 aims to reduce inequalities in income, age, gender, disability, race, ethnicity, origin, and religion within and between countries. It requires a commitment between countries to ensure safe migration and the mobility of people. The indicators monitor income distribution within countries, the risk-of-poverty gap and disparities between countries, employment rate, migration, and social inclusion.

SDG 11—Sustainable cities and communities. SDG 11 aims to renew and plan cities and human settlements, guaranteeing access to basic services, housing, transport, and public green spaces through the efficient use of resources and a low environmental impact. Monitoring includes the quality of life in cities, sustainable mobility, and environmental impacts.

SDG 12—Responsible consumption and production. SDG 12 aims to develop sustainable actions and practices by businesses, public bodies, and consumers through advanced technologies, resource efficiency, and waste reduction. Monitoring covers the decoupling of environmental pressures from economic growth, green economy, waste generation, and management.

SDG 13—Climate Action. SDG 13 aims to limit global warming, strengthen climate resilience and enhance countries' capacity to adapt, ensuring support for the least developed countries. Monitoring indicators consider climate mitigation, climate impacts and adaptation, and financing climate action.

SDG 14—Life below water. SDG 14 aims to protect the sustainability of oceans by reducing marine pollution and ocean acidification. It aims to eliminate overfishing and protect and conserve marine and coastal ecosystems. SDG 14 indicators include monitoring trends in ocean health, marine conservation, and sustainable fisheries.

SDG 15—Life on earth. SDG 15 aims to protect and restore the sustainability of terrestrial ecosystems through sustainable forest management, limiting deforestation and desertification, restoring biodiversity, and safeguarding endangered species. Monitoring covers trends in ecosystem status, land degradation, and biodiversity.

SDG 16—Peace, justice, and strong institutions. SDG 16 promotes peaceful and inclusive societies based on respect for human rights, protection of the most vulnerable, and good governance practices. Transparency, efficiency, and accountability are central elements of the objective. Indicators monitor peace and personal security, access to justice, and trust in institutions.

SDG 17—Partnership for goals. SDG 17 promotes a global partnership for sustainable development and fair trade. It aims to ensure macroeconomic stability, financial support for developing countries, and access to science, digital information and communication technology. Monitoring includes global partnership, financial governance, and access to technology.

Clearly, all the SDGs have a direct or indirect impact on the firm's structure and its management, also in relation to the sector to which it belongs and its position in the supply chain, in the entrepreneurial ecosystem and in a specific territorial context. Given the complexity of the SDGs and their implications in various sectors of the economy, this book focuses on SDG 9-*Industry, innovation, and infrastructure,* SDG 11—*Sustainable cities and communities,* and SDG 12—*Responsible consumption and production.* They specifically address the aspects of the product, service, and territory in which the firm is directly involved in managing sustainability. Furthermore, SDGs 9, 11, and 12 are among the priority interventions that the European Commission has highlighted within the Green Deal. Fig. 2.2 shows the European Commission Priorities, divided into six headline ambitions: European Green Deal, Economy that works for people, Europe fit for the digital age, European way of life, Stronger Europe in the world, and European democracy.

The European Green Deal, approved in 2019 (European Commission, 2019), has given a change to EU policies on climate neutrality, reduction of net greenhouse gas emissions, and impact on the most vulnerable sections of the population. It represents a transformative and integrative policy aimed at achieving a modern, efficient, and competitive economy in resources use, managing climate and environmental challenges, and governing a just and inclusive transition (Fig. 2.3).

The European Green Deal policies are reinforced by a series of initiatives and strategies promoted by the European Commission. Among them, the 2020 *Circular Economy Action Plan* introduces measures related to the entire life cycle of products, design, and production for a circular economy, with the goal of securing and maintaining resources within the EU economy. The *EU Bioeconomy Strategy* provides a framework to foster the development of transformative innovations affecting the use of biological resources and biomass supply. The purpose of a sustainable circular bioeconomy is to

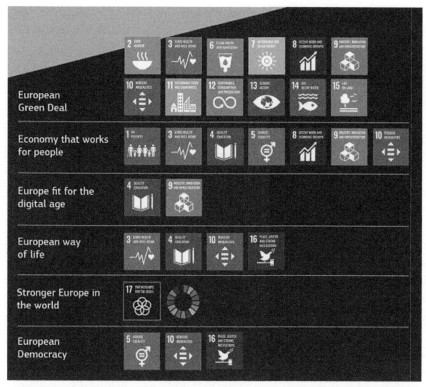

Figure 2.2 European Commission priorities. *Source: From Eurostat (2023, p. 22.).*

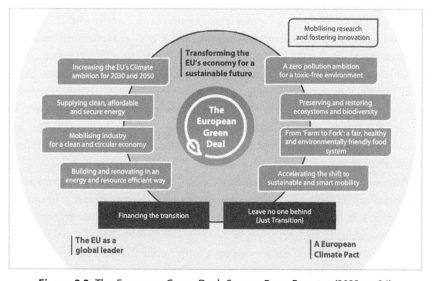

Figure 2.3 The European Green Deal. *Source: From Eurostat (2023, p. 24).*

mitigate climate change through renewable products and energy and the Farm to Fork approach. This strategy is based on fair, healthy, and environmentally friendly food systems.

The *EU Biodiversity Strategy*, introduced in 2020 (European Union, 2020), contains actions and measures to create a broad network of land and sea-protected areas and to implement a nature and biodiversity restoration plan.

The *2030 Climate Target Plan* (European Commission, 2020) includes a reduction in greenhouse gas emissions and a responsible path to achieving climate neutrality by 2015. The proposals cover economic transformation, building renovation, sustainable transport, renewable energy, and enhancement in global climate action. The *Council Recommendation on ensuring a fair transition toward climate neutrality* aims at ensuring a climate-neutral, fair, and environmentally sustainable economy. The 2020 *Sustainable and Smart Mobility Strategy* intends to transform the EU transport system in a green, digital, and resilient key. The *2021 Zero Pollution Action Plan* aims to reduce air, water, and soil pollution to protect health and natural ecosystems.

The *European Pillar of Social Rights Action Plan* (European Commission, 2021a) foresees active involvement of EU Member States, social partners, and civil society to monitor progress achieved in employment, skills development, and poverty reduction. The 2021 EU's new *Industrial Strategy* promotes the double transition toward a green and digital economy to favor a more inclusive and resilient industry.

The 2023 *Green Deal Industrial Plan* (European Commission, 2023) focuses on improving the competitiveness of the EU's zero-emission industry and creating more favorable conditions for the development and production of net-zero technologies and products. The 8th Environment Action Program (EAP), introduced in 2022, extends Member States' commitment to environmental action to 2050 and defines six priority objectives such as climate neutrality, climate adaptation, circular economy, zero pollution, biodiversity protection and restoration, and the reduction of environmental and climatic pressures.

Aligned with Agenda 2030 and the SDGs, the *Global Gateway* (European Commission, 2021b) is the EU strategy to promote smart, clean, and secure connections in the digital, energy, and transport sectors. It also strengthens health, education, and research systems according to the Team Europe approach, launched in 2020 to support partner countries in the fight against the COVID-19 pandemic and its consequences.

These programs are reinforced by the EU's *economic, social, and territorial cohesion policy* (European Union, 2021), supported by various social funds, including European Regional Development Fund—ERDF, European Social Fund+ (ESF+), Cohesion Fund, Just Transition Fund— JTF. They aim to correct imbalances between countries and regions.

The EU's *Horizon Europe* (European Commission, 2021c) research and innovation program supports researchers in developing solutions for the transition to a green, digital, and resilient Europe.

National Recovery and Resilience Plans (RRPs) (European Commission, 2021d) are structured around six thematic pillars: green transition; digital transformation; economic cohesion, productivity, and competitiveness; social and territorial cohesion; health, economic, social, and institutional resilience; and policies for the next generation. They address the four dimensions of competitive sustainability outlined in the economic policy agenda called the *Annual Sustainable Growth Survey* (ASGS) 2023, such as environmental sustainability, productivity, equity, and macroeconomic stability. Each dimension corresponds to different SDGs and a reform and investment plan in the PRR to support the green and digital transition.

2. SDG 9: Industry, innovation, and infrastructure

What it is: SDG 9 entails building resilient infrastructure and promoting innovation and equitable, responsible, and sustainable industrialization. SDG 9 goal is to foster economic growth, employment, and social welfare, through inclusive and sustainable industrialization. Clean and environmentally friendly technologies and industrial processes are particularly supported in developing countries. SDG 9 also aims to facilitate the access of small and medium-sized enterprises to financial services under advantageous conditions, to increase integration into markets and value chains.

What it consists of: with the increase in population worldwide, it is necessary to build infrastructure based on sustainability and quality criteria. Infrastructure includes roads, shipping links, access to water, electricity, and the Internet. SDG 9 requires companies to adopt sustainable business models where production processes make use of cutting-edge technologies and circular economy systems. Accessibility is also understood as people's ability to access medical care and educational/training opportunities.

How to decline: The actions underlying SDG 9 are:
- Build infrastructures, including regional and cross-border ones, of quality, safe, sustainable, and resilient to promote equitable access and the

well-being of individuals and economic development. Specifically, provide financial support for the technological development and construction of sustainable infrastructure in African countries, least developed countries, landlocked countries, and small Island developing States.

- Strengthen industry in an inclusive and sustainable way and increase the level of employment and gross domestic product, especially in less developed countries.
- Facilitate access to financial services for small businesses, especially those located in developing countries.
- Increase efficiency in the use of resources through the modernization of structures and the adoption of cutting-edge technologies and ecological industrial processes.
- Boost scientific research, technological skills, and employment in the R&D sector, particularly in developing countries, also through an environmental policy favorable to industrial diversification.
- Ensure access to ICT technologies with universal and affordable internet for the least developed countries.

What is the role of firms: SDG9 favors green industrial processes that include small and medium-sized enterprises in its value chain. In other words, businesses are called upon to contribute to the development of communication infrastructures and technologies in the area in which they operate.

How to monitor firms' contribution: The main indicators for monitoring the contribution of firms are total expenditure and investments devoted to environmental protection; development and impact of investments in infrastructure and services; size, typology, and impact of technological heritage; direct economic value generated and distributed.

SDG9 trend to date: Based on the Eurostat publication "Sustainable development in the European Union—Monitoring report on progress toward the SDGs in an EU context—2023 edition" (p. 13), SDG 9 shows a positive trend in most of its indicators, including patent applications to the European Patent Office, the share of R&D workers in the labor force, and the share of young people with tertiary education. The R&D and innovation indicator has experienced moderate growth due to the size of R&D spending. The values of the sustainability transformation of the EU industrial sector and those of atmospheric emissions from the manufacturing sector have shown significant improvement in the total economy. Sustainable infrastructures have experienced moderate development, with the exception of sustainable passenger and freight transport, which recorded below-average ecological values.

3. SDG 11: Sustainable cities and communities

What it is: SDG 11 means making cities and human settlements inclusive, safe, resilient, and sustainable. It aims to reduce the per capita pollution produced by cities, with particular reference to air quality and waste management. SDG 11 supports inclusive and sustainable urban development and promotes participatory and integrated settlement planning. This means ensuring access to safe and inclusive green surfaces and public spaces for women and children, the elderly, and people with disabilities.

What it consists of: The ever-growing phenomenon of urbanization and the expansion of cities has favored social and economic progress and, at the same time, has caused situations of degradation and poverty linked to inefficient management of natural resources and the lack of funds to ensure essential services and adequate housing facilities. SDG 11 is aimed at transforming urban centers into sustainable cities through expanded access to adequate, affordable, and safe housing, basic services, and means of transport for all population groups. The design of the green cities of the future involves a reduction of environmental impacts, the enhancement of green areas and urban suburbs, and the preservation of the common artistic and cultural heritage.

How to decline: The actions underlying SDG 11 are:
- Ensure access to affordable, adequate, and safe housing and basic services, regenerate poor neighborhoods, and support least-developed countries in the construction of sustainable and resilient buildings using local materials, including technical and financial assistance.
- Promote a safe, convenient, accessible, and sustainable transport system, investing in road safety and public transport, with particular attention to the needs of the most vulnerable subjects.
- Enhance inclusive and sustainable urbanization and human settlement planning in a participatory and inclusive way.
- Protect and safeguard the historical, cultural, and natural heritage of the planet.
- Significantly reduce the number of deaths and sick people, limiting economic losses caused by natural disasters and paying particular attention to the protection of the poor and vulnerable people.
- Implement integrated policies and action plans for resource efficiency, climate change mitigation and adaptation, and disaster resilience, through holistic disaster risk management at all levels, in line with the Sendai Framework for Risk Reduction of Disasters 2015—30.

- Improve air quality and waste management and create safe and accessible green and public spaces.
- Enhance the economic, social, and environmental links between urban, periurban, and rural areas through national and regional development planning.

What is the role of firms: SDG11 assumes that firms adopt production processes in compliance with the protection of the cultural and natural heritage of the place where they are located and invest in green mobility and in projects/initiatives for the protection from environmental disasters. While various sectors are involved in achieving SDG 11, those directly affected are those in the construction sector.

How to monitor firms' contribution: Specific indicators to monitor the contribution of firms in achieving SDG 11 are the number, type, and impact of the promotion of the use of sustainable means of transport; quantity and impact of investments in infrastructure and green transport services; number, type, and impact of contingency measures and programs focused on the prevention and management of disasters and emergencies; number and percentage of people who have moved locally due to noise deriving from company activities.

SDG11 trend to date: Based on the Eurostat publication "Sustainable development in the European Union—Monitoring report on progress toward the SDGs in an EU context—2023 edition" (p. 14), SDG 11 shows significant improvements in the quality of life in city and community. In particular, serious housing problems, premature deaths due to exposure to fine particles and the rate of crime, violence, and vandalism in neighborhoods have shown a decreasing trend. Perceived exposure to noise decreased. Conversely, sustainable mobility and environmental impacts have had more limited developments due to changes in mobility patterns due to the COVID-19 pandemic. Road deaths have increased, also due to a sharp decline in the use of public passenger transport (buses and trains). Land consumption has increased rapidly due to the increase in settlement areas in absolute and per capita terms. Finally, the increase in the recycling rate of EU municipal waste has slowed down in recent years.

4. SDG 12: Ensure sustainable consumptions and production patterns

What it is: SDG 12 means guaranteeing the well-being of the population through access to water, energy, and food, reducing the waste of natural resources. Goal 12 promotes sustainable consumption and production models through an environmentally friendly approach by limiting the use of chemicals. The goal aims at a significant reduction of waste and food waste by promoting sustainable business management and public procurement based on sustainability criteria.

What it consists of: Traditional production systems and consumption models involve a considerable waste of resources and damage to ecosystems. An efficient use of natural resources and a fair redistribution of electricity, drinking water, and food among the population represent the basis for changing the current model of production and consumption. SDG 12 focuses on the responsible and efficient management of natural resources, which is achieved with processes to reduce food waste, use of eco-sustainable chemical products, and reduce waste. SDG12 is aimed at companies to start sustainable production processes, at people to stimulate green consumption and practices, and at institutions to activate a series of measures at the regulatory level.

How to decline: The actions underlying SDG 9 are:
- Implement the Ten Year Framework of Programs for Sustainable Consumption and Production, through the participation of developed countries, which will have to take into account the opportunities and capacities of developing countries.
- Promote sustainable management and efficient use of natural resources, reduce global food waste per capita in sales, and minimize food losses along production and supply chains.
- Reduce waste production through prevention, reduction, recycling, and reuse.
- Adopt a circular economy system for the disposal of chemicals and waste throughout their life cycle, reducing their release into the air, water, and soil to minimize their negative impact on human health and the environment.
- Support companies, especially large multinationals, to adopt sustainable practices and provide sustainability information in their annual reporting and communication tools.
- In public procurement, promote sustainable criteria and practices in accordance with national policies and regulations.
- Raise awareness of sustainable development and the importance of adopting a lifestyle in harmony with nature among the population.
- Strengthen the scientific and technological capabilities of developing countries to encourage a shift towards more sustainable consumption and production models.

- Adopt tools to monitor the impacts of sustainable development on job creation and local culture and products promotion.
- Limit the subsidies foreseen for the use of fossil fuels that encourage waste by restructuring the taxation systems, considering the conditions of developing countries and potential negative effects on the poor and local communities.

What the role of firms: SDG12 assumes that firms, especially those operating in the food, textile, tourism, and consumer goods sectors, make a significant contribution to achieving the goal, for example, through the use of recyclable and biodegradable materials in their processes production and the withdrawal from the market of products with high energy consumption or high environmental impact.

How to monitor firms' contribution: Specific indicators for monitoring firms activities include: energy consumption reductions in production and distribution processes; type and number of sustainability, quality, and ethics certificates; percentage of recycled materials used; environmental impact rate of products and services; and quantity and type of actions for the diffusion of responsible consumption practices.

SDG12 trend to date: Based on the Eurostat publication "Sustainable development in the European Union—Monitoring report on progress toward the SDGs in an EU context—2023 edition" (p. 13), SDG 12 shows a contrasting trend in monitoring indicators. The EU's material footprint decreased temporarily in 2020 due to reduced economic activity during the pandemic period. Domestic material consumption data indicate an increase in the use of resources and the consumption of hazardous chemicals. The average efficiency indicator of carbon dioxide (CO_2) emissions of new passenger cars has improved. Total production and waste management show contrasting values in the pre- and post-COVID period. The EU's circular material utilization rate has remained below 12% in recent years. Lastly, the environmental goods and services sector had growth in gross value added above the sector average.

References

European Commission (2019). (2019). The European green deal. https://commission.europa.eu/strategy-and-policy/priorities-2019-2024/european-green-deal_en. (Accessed 28 July 2023).

European Commission (2020). (2020). 2030. Climate target plan. https://ec.europa.eu/info/law/better-regulation/have-your-say/initiatives/12265-2030-Climate-Target-Plan_en. (Accessed 28 July 2023).

European Commission (2021a). (2021). European pillar of social rights action plan. https://ec.europa.eu/social/main.jsp?catId=1607&langId=en. (Accessed 28 July 2023).

European Commission (2021b). (2021). Global gateway strategy. https://eur-lex.europa.eu/legal-content/EN/TXT/?uri=CELEX%3A52021JC0030. (Accessed 28 July 2023).

European Commission (2021c). (2021). Horizon Europe. https://commission.europa.eu/funding-tenders/find-funding/eu-funding-programmes/horizon-europe_en. (Accessed 28 July 2023).

European Commission (2021d). (2021). National recovery and resilience plans. https://commission.europa.eu/business-economy-euro/economic-recovery/recovery-and-resilience-facility_en. (Accessed 28 July 2023).

European Commission (2023). 2023. Green deal industrial plan. https://commission.europa.eu/system/files/2023-02/COM_2023_62_2_EN_ACT_A%20Green%20Deal%20Industrial%20Plan%20for%20the%20Net-Zero%20Age.pdf. (Accessed 28 July 2023).

Eurostat. (2023). Sustainable development in the European Union. Monitoring report on progress towards the SDGs in an EU context. 2023 edition, p. 21. https://ec.europa.eu/eurostat/web/products-flagship-publications/w/ks-04-23-184. (Accessed 28 July 2023).

European Union (2020). (2020). EU biodiversity strategy. https://eur-lex.europa.eu/EN/legal-content/summary/biodiversity-strategy-for-2020.html. (Accessed 28 July 2023).

European Union (2021). (2021). Economic, social and territorial cohesion policy. https://www.europarl.europa.eu/factsheets/en/sheet/93/economic-social-and-territorial-cohesion#:~:text=Context,and%20protection%20of%20the%20environment. (Accessed 28 July 2023).

CHAPTER THREE

SDG indicators set for sustainability assessment

1. UN SDG indicators set

The United Nations Statistical Commission, with the support of the Inter-Agency and Expert Group on SDG Indicators (IAEG-SDGs), defined the global indicator framework for the Sustainable Development Goals in 2017. The General Assembly Resolution on the work of the Statistical Commission on the 2030 Agenda for Sustainable Development (A/RES/71/313) refined the modalities for adoption, establishing that the indicators will be updated annually and complemented by developed indicators at regional and national level proposed by the Member States (Resolution Adopted by the General Assembly, 2017).

The official list of indicators includes the Global Indicator Framework (A/RES/71/313), refinements agreed by the Statistical Commission in 2018 (E/CN.3/2018/2, Annex II) and in 2019 (E/CN.3/2019/2, Annex II), changes from the 2020 version (E/CN.3/2020/2, Annex II) and related improvements (E/CN.3/2020/2, Annex III), 2021 refinements (E/CN.3/2021/2, Annex), sanctions (E/CN.3/2022/2, Annex I) and decisions (53/101) of the United Nations Statistical Commission (E/2022/24-E/CN.3/2022/41). The total number of SDG indicators is 248, 13 of which are repeated several times (United Nations, 2023). Ultimately, the global framework of SDG indicators is made up of 231 specific indicators.

The Sustainable Development Goals Report 2023 illustrates the progress of the objectives compared to the values of the SDG indicators (Sustainable Development Goals Report, 2023). The measurement of progress achieved depends on the quality and availability of the data and information underlying the measurement of the indicators. By 2023, progress towards the 17 objectives displays an uneven trend, categorized into four areas: on track or target met; fair progress, but acceleration needed; stagnation or regression; and insufficient data.

Being a Sustainable Firm
ISBN: 978-0-443-14062-4
https://doi.org/10.1016/B978-0-443-14062-4.00001-5

As shown in Fig. 3.1, SDGs 12, 14, 9, and 15 are among those with the highest percentage on track or target met; G1, G6, G8, and G13 are among the goals that, despite not reaching a target, are in a better position for fair progress, but acceleration is needed. G14, G8, and G2 are the indicators with the highest percentage of stagnation or regression. G11, G16, and G5 objectives have a higher percentage of insufficient data.

The data comparability at an international level has been improved by the increase in the number of indicators included in the SDGs global database, which moved from 115 to 225 in the period 2016—2023. The methodological development of SDG indicators has also seen an improvement in terms of the adoption of consolidated methodologies or internationally established standards. This result is crucial for ensuring the comparability, reliability, consolidation, and harmonization of the methodologies. A clearer and more comparable indicator framework provides a solid basis for monitoring SDG performance and comparing SDG progress across countries. Fig. 3.2 shows a picture of SDG progress and data monitoring strides. It highlights that only 15% of SDG progress is on track while a significant increase in data monitoring is registered (from 330.00 in 2016 to 2.7 millions of data in 2023).

Progress assessment for the 17 Goals based on assessed targets, 2023 or latest data (percentage)

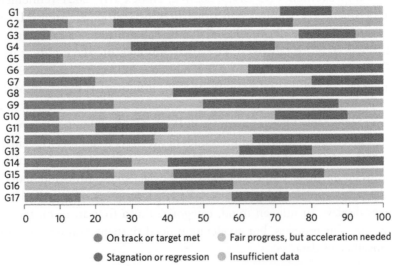

Figure 3.1 Progress assessment for the 17 goals based on assessed targets, 2023 or latest data (percentage). *From The-Sustainable-Development-Goals-Report-2023: 8.*

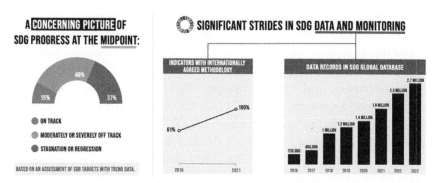

Figure 3.2 A picture of UN SDG progress and data monitoring. *From The-Sustainable-Development-Goals-Report-2023: 11.*

The considerable demand for data necessitates the utilization of a multiplicity of sources. Alongside traditional sources, such as phone-based interviews, surveys and web-based data collection and official documents, innovative sources based on modern technologies, and inclusive approaches are emerging. Non traditional data sources include internal administrative documents, satellite images and earth observation, scanner/credit card data, mobile phone data, social media, and citizen-generated data. Integration of multiple data sources, data timeliness, and disaggregation of indicators by microareas are essential components in data production. The "small area estimation" approach is becoming increasingly prevalent for measuring SDG indicators related to social protection, health, education, and employment.

To improve the monitoring of the targets, countries are developing a "whole-of-society" approach, engaging in partnerships with a wide range of stakeholders. This involves establishing partnerships at national and international levels, also through collaborations between the competent offices and ministries. The public sector is the main partner of national statistical offices, followed by academia, the private sector, and civil society organizations. A small percentage (13%) of countries has yet to formalize partnership agreements, while at an international level, the SDG monitoring process is promoting the collaboration among bodies, experts, and stakeholders from different communities capable of providing data at local, national, regional, and international. A higher openness, accessibility, and effective use of data have been promoted by several countries to amplify their impact, conduct more in-depth analysis, and promote transparency and accountability. The accuracy of the data supports policymakers in defining policies and allocating funds as well as promoting significant social changes, especially in low- and lower-middle-income countries.

For the implementation of the core Sustainable Development Goal indicators in different organizations from different countries (such as listed companies, SMEs, and Family business), the United Nations Conference on Trade and Development (UNCTAD) has developed a set of indicators covering the economic, social, and institutional dimensions to support companies in SDG reporting . Below are the four areas in which the key indicators are identified in the UNCTAD Guide (2022) according to corporate reporting principles and practices (UNCTAD, 2022).

Economic area indicators:
- Revenue
- Value added (gross value added, GVA)
- Net value added (NVA)
- Taxes and other payments to the Government
- Green investment
- Community investment
- Expenditures on research and development
- Share of local procurement

Social area indicators
- Share of women in managerial positions
- Hours of employee training
- Expenditures on employee training
- Employee wages and benefits
- Expenditures on employee health and safety
- Incidence rate of occupational injuries
- Share of employees covered by collective agreements

Environmental area indicators
- Water recycling and reuse
- Water use efficiency
- Water stress
- Waste generation
- Waste reused, remanufactured and recycled
- Hazardous waste generation
- Greenhouse gas emissions
- Ozone-depleting substances and chemicals
- Share of renewable energy
- Energy efficiency
- Land used adjacent to biodiversity sensitive areas

Institutional area indicators

- Board meetings and attendance
- Share of female board members
- Board members by age range
- Audit committee meetings and attendance
- Compensation per board member
- Corruption incidence
- Management training on anticorruption

The division of indicators into specific areas provides practical information to measure the SDGs with coherent and comparable indicators across different countries to facilitate the monitoring of the 2030 Agenda achievement. The key indicators for areas are effective tools to support governments and the private sector in sustainability detection, evaluation, and reporting.

2. EU SDG indicators set

At the EU level, the indicators monitoring to achieve the 2030 Agenda evaluate the EU's progress toward the goals based on the official EU SDG indicator set. While this set of indicators is aligned with the UN list of global indicators, it is not identical, as it focuses on monitoring community policies and SDG progress in the European context. Specifically, 68 of the EU SDG indicators are aligned with the UN SDG indicators. In addition, preference is given to indicators that have strong connections to EU policy initiatives and are part of a high-level scoreboard for social rights.

Starting with the global SDG indicator list, which includes 232 indicators, the United Nations Economic Commission for Europe (UNECE) selected 80 indicators based on data availability, regional political relevance from an EU perspective, national coverage, up-to-datedness, and data quality (KnowSDGs). The resulting UNECE SDG dashboard guides statistical bodies and offices in measuring the achievement of the 2030 Agenda objectives across different areas, including national coordination, reporting on global indicators, monitoring progress at various levels, quality assurance, and communication.

Eurostat annually prepares monitoring reports on the SDGs in the EU context considering the trends over the last 5 years. The report offers ongoing monitoring of progress, highlighting the cross-cutting nature and connections between the SDGs. It supports the coordination of sustainability policies at the European level among Member States. The 2023 Monitoring (Eurostat, 2023) report shows that considerable progress has been made toward socioeconomic objectives, while advancements toward environmental goals have been limited. The evaluation method considers the tendency of the objective with respect to the goal, establishing whether an indicator has moved closer or further away from it and the speed of this movement. In particular, significant progress has been made in achieving SDG 8 (decent work and economic growth), SDG 1 (poverty reduction), and SDG 5 (gender equality). Good progress was achieved for SDG 10 (reduction of inequalities), SDG 4 (quality education), SDG 16 (peace, justice, security, and strong institutions), SDG 3 (good health and well-being), and SDG 9 (innovation and infrastructure). The remaining objectives have registered moderate improvements (SDG 12, 11, 14, 2, 6, 7). The EU is implementing policies and measures to improve the three climate objectives having experienced moderate movement away, SDG 13 (climate action), SDG 15 (life on earth), and SDG 17 (global partnerships). Fig. 3.3 gives an overview of the EU short-term progress (over the past 5 years) towards the SDGs in 2023.

The EU SDG indicator series (Eurostat Unit, 2023) evaluates the 17 SDGs according to the social, economic, environmental, and institutional dimensions of sustainability. Each SDG is measured by five or six headline indicators selected in accordance with their aims and ambitions (see the Appendix at the end of the chapter).

The EU SDG indicators set exhibits the following characteristics:
- Lens used to select each of 17 lenses
- Indicator code used in the Eurostat database
- MPI multipurpose indicator, used to indicate the objective(s) to which the indicator is assigned also for monitoring purposes
- Name and unit of measurement of the indicator presented in the Eurostat database
- Frequency of data collection divided into categories: every year; every 2 years; every 3 years; every >3 years; a-periodic
- Geographic coverage including EU aggregate and all Member States, plus other countries
- Data source including data collection and surveys on which the indicator's data is based
- Data providers, which include services or institutions disseminating the underlying indicator or data.

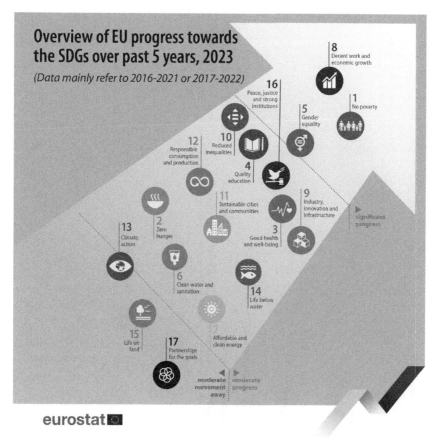

Figure 3.3 Overview of EU progress toward the SDGs over the past 5 years, 2023 (Data mainly refer to 2016–2021 or 2017–2022). *From Sustainable development in the European Union. Monitoring report on progress toward the SDGs in an EU context 2023 edition: 11.*

Out of 100 indicators, 33 are deemed "multipurpose," serving to monitor more than one objective and highlighting their connection. The EU SDG indicator set is subject to periodic revisions in relation to new policy developments and may be updated following the introduction of new methodologies and data sources.

Table 3.1 summarizes the characteristics of the EU SDG indicator set 2023 as described earlier, highlighting the number of main and multipurpose indicators. Notably, EU SDGs 4, 7, and 14 lack multipurpose indicators. The frequency of data collection ranges from 3 to 6 times a year. The last columns indicate the alignment of each indicator with the UN list and the number of indicators either replaced or integrated.

Table 3.1 EU SDG indicator set 2023.

Characteristics of EU SDG indicator set 2023

Main features of the EU SDG indicator set
Structured along
i. the 17 global SDGs
Limited to 6
ii. indicators per goal
Multipurpose indicators used to complement monitoring of each goal
iii.

	Indicators per goal		Selected indicators, of which			Indicators replaced	Indicators adjusted	Indicators on hold
	Main	+ MPI	Annual frequency	Provider Eurostat	In UN list			
1 No poverty	6	3	6	6	5			1
2 Zero hunger	5	3	4	4	1			2
3 Good health and well-being	6	5	5	5	3			2
4 Quality education	6		4	5	5		1	
5 Gender equality	6	2	5	3	5		1	
6 Clean water and sanitation	6	1	4	2	5		1	2
7 Affordable and clean energy	6		6	6	3			
8 Decent work and economic growth	6	3	6	6	4			1

9 Industry, innovation, and infrastructure	6	3	6	5	5			
10 Reduced inequalities	6	5	6	6	4			1
11 Sustainable cities and communities	6	3	3	4	5	1	2	2
12 Responsible consumption and production	6	1	5	5	2	1	1	3
13 Climate action	5	2	5		3			2
14 Life below water	6		5		4			
15 Life on land	6	2	3	1	5			3
16 Peace, justice, and strong institutions	6		5	3	5		1	
17 Partnership for the goals	6		6	3	4			
Total	**100**	**33**	**84**	**64**	**68**	**2**	**9**	**19**
Compared to 2022 version	−1	+2	−5	+1	+1	—	—	—

From Eurostat May 23, 2023 – EU SDG indicators 2023.

3. ISTAT SDG indicators set in Italy

The Italian National Statistics Institute(ISTAT) plays the role of coordination and production of statistical information useful for monitoring the 2030 Agenda results in Italy. ISTAT works in collaboration with several bodies to carry out an interinstitutional comparison. They include Higher Institute for Environmental Protection and Research, Ministry of Economy and Finance, Energy Services Manager, Ministry of Justice, Higher Institute of Health, Ministry of the Interior, Invalsi—National Institute for the Evaluation of Education and Training System, Ministry of University Education and Research, ENEA—National Agency for New Technologies, Energy and Sustainable Economic Development, Ministry of Health, INGV—National Institute of Geophysics and Volcanology, ASviS—Italian Alliance for Sustainable Development, Ministry of the Environment and Protection of Land and Sea, Consob—National Commission for Companies and the Stock Exchange, Ministry of Foreign Affairs and International Cooperation, Cresme—Economic and Social Research Center of Construction Market.

ISTAT–SDGs System has been continuously evolving since 2016 in the production of new measures, updates to previous measures, and methodological advancements within the activities of the Inter-Agency and Expert Group (SDGs UN-IAEG-SDGs). The ISTAT Report on the 2023 Sustainable Development Goals (SDGs) includes 372 statistical measures, of which 342 are unique, that is, associated with a single goal and 30 are multipurpose. These measures are linked to 139 indicators from the set proposed by UN-IAEG-SDGs. Furthermore, 223 out of a total of 372 statistical measures have been updated, and 5 new measures have been added.

The ISTAT-SDGs statistical measures have points of contact with the Fair and Sustainable Well-being (BES) indicators contained in the Economic and Financial Document (DEF).

Upon the initiative of the Ministry of the Environment and Protection of Land and Sea, a working group was established to develop a smaller and more representative group of indicators to effectively monitor the Goals. The technical table, which included representatives from the Ministry of the Environment and Protection of Land and Sea, the Ministry of Economy and Finance, the Ministry of Foreign Affairs and International Cooperation, ISPRA, and ISTAT, defined and agreed on the criteria (such as parsimony, feasibility, timeliness, extent and frequency of time series, sensitivity to public policies, and territorial dimension) and methodologies.

Currently, the SDGs system shares 62 indicators with the BES system to ensure integration with the evaluation of public policies. Fig. 3.4 shows the

Bes		SDGs	
1. Health	4 indicators	4 in Goal 3	
2. Education and training	8 indicators	7 in Goal 4 1 in Goal 8	
3. Work and life balance	10 indicators	2 in Goal 5 8 in Goal 8	
4. Economic well-being (a)	7 indicators	5 in Goal 1 3 in Goal 10	
5. Social relationships			
6. Politics and Institutions (a)	8 indicators	4 in Goal 5 5 in Goal 16	
7. Security	3 indicators	1 in Goal 5 2 in Goal 16	
8. Subjective well-being			
9. Landscape and cultural heritage	2 indicators	1 in Goal 11 1 in Goal 13	
10. Environment (b)	11 indicators	1 in Goal 1 2 in Goal 6 1 in Goal 7 1 in Goal 8 3 in Goal 11 2 in Goal 12 2 in Goal 13 1 in Goal 14 2 in Goal 15	
11. Innovation, research and creativity	3 indicators	3 in Goal 9	
12. Quality of services(a)	8 indicators	2 in Goal 1 3 in Goal 3 1 in Goal 6 2 in Goal 11 1 in Goal 16	

(a) 1 indicator is in more than one Goal.
(b) 4 indicators are in more than one Goal.

Figure 3.4 Istat-SDGs statistical measures and BES indicators, by BES domain and SDGs goal. *From 2022 SDGsReport. Statistical information for 2023 Agenda in Italy: 10.*

correspondence between BES indicators, ISTAT–SDGs statistical measures, and SDGs objective (ISTAT, 2022).

The temporal trends of the measures in the last year (2021 or 2022) compared to past 10 years show positive signs of improvement. Almost 60% of the measures recorded an improvement, approximately 21% remained stable, while approximately 20% worsened. Goals 5 (gender equality), 7 (affordable and clean energy), 8 (decent work and economic growth), 12 (responsible consumption and production), 16 (peace, justice, and strong institutions), and 17 (partnership) have recorded a higher level of improvement. Goals 2 (defeat hunger), 4 (quality education), 11 (sustainable cities and communities), and 13 (fight against climate change) have instead worsened in more than a third of the indicators. Fig. 3.5 shows the temporal evolution of statistical measures for the SDG calculation. The number of measures used for the calculation is indicated in brackets next to each goal.

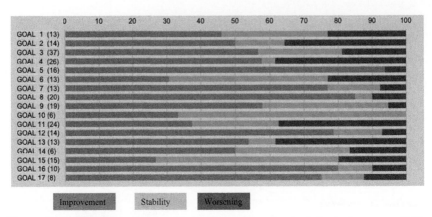

Figure 3.5 Temporal evolution of statistical measures in Italy: last year available compared to the previous 10 years, for goal. *From Rapporto SDGs (2023). Informazioni statistiche per l'Agenda 2030 in Italia: 12.*

To monitor the objective of reducing disparities between regions over time, the SDGs report uses measures of interregional convergence over time. Overall, over the last 10 years, almost half of the 159 statistical measures used indicate a convergence between the regions, especially for goals 9 (infrastructure) and 17 (partnerships) linked to digitalization, research, and development. Territorial inequality reduction is also observed for Goals 4 (education), 10 (reduce inequalities), and 13 (climate change) in the area of student skills, a more equitable distribution of income, and reduced exposure to the risk of natural disasters. Conversely, a greater divergence is recorded for Goals 7 (clean and accessible energy) and 11 (sustainable cities and communities) due to the gap in the share of energy consumption from renewable sources and access to public transport. The following Fig. 3.6 shows the combination of the temporal evolution of statistical measures with respect to the objectives of the 2030 Agenda (improvement vs. stability/worsening) and the convergence between regions (convergence vs. stability/divergence).

Fig. 3.7 illustrates Italy's positioning, in terms of SDG objectives compared to the European Union.

In particular, the yellow quadrant on the left indicates that Italy is moving away from the indicated target, but the status is better than that of the EU. The green quadrant indicates that Italy is progressing toward these sustainable development goals and its status is better than that of the EU. The red quadrant representation indicates that Italy is moving away from the

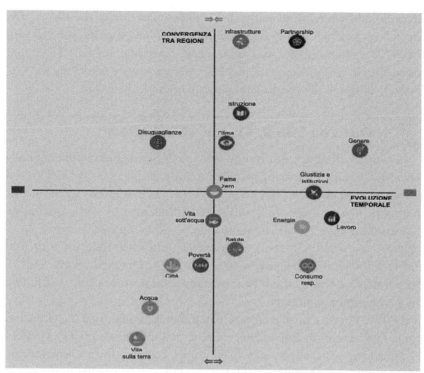

Figure 3.6 Goals according to temporal evolution and convergence between regions: last year available compared to the previous 10 years. *From Rapporto SDGs (2023). Informazioni statistiche per l'Agenda 2030 in Italia: 14.*

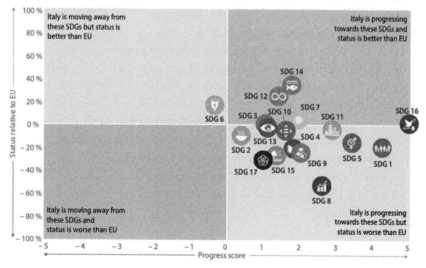

Source: Eurostat

Figure 3.7 Italy SDG's status and progress. *From 2022 SDGs Report. Statistical information for 2023 Agenda in Italy: 324.*

objectives shown, with its status being worse than that of the EU. Lastly, the yellow quadrant on the right highlights that Italy is progressing toward the SDGs represented, but its status is worse than that of the EU. The figure shows that Italy is making considerable progress toward achieving the objectives, although the results fall below than the EU average.

4. Focus on SDG 9, 11, and 12 indicators

Goal 9—Industry, innovation, and infrastructure means building resilient infrastructure, promoting inclusive and sustainable industrialization, and encouraging innovation (SDG 09). SDG 9 is made up of 8 targets and 12 indicators, in the areas of sustainable infrastructure (9.1, 9.4), sustainable industrialization (9.2, 9.4), strengthening access to financial services (9.3), scientific research (9.5), tools for implementing sustainable infrastructure and technological development (9.a, 9.b), and improved access to ICT (9.c). Table 3.2 reports the UN targets and indicators for goal 9, including the multipurpose indicators.

GOAL 11—Sustainable cities and communities mean making cities and human settlements inclusive, safe, resilient, and sustainable (SDG 11). SDG 11 is made up of 10 targets and 15 indicators, covering the areas of sustainable housing and urbanization (11.1, 11.3, 11.6, 11.7), specific measures on transport system (11.2), cultural and natural heritage (10.4), human and economic disaster loss reduction (11.5), tools for implementing integrated national and regional policies and plans (11.a, 11.b), and sustainable buildings (11.c). Table 3.3 reports the UN targets and indicators for goal 11, including the multipurpose indicators.

Goal 12—Responsible production and consumption means ensuring sustainable consumption and production patterns (SDG 12). SDG 12 is made up of 11 targets and 13 indicators in the field of sustainable production and consumption patterns (12.1, 12.2), waste reduction (12.3, 12.4, 12.5), availability of relevant sustainability information both at company's level (12.6, 12.7) and people's level (12.8), tools for implementing promotion and monitoring of sustainable patterns (12.a, 12.b), and market distortions reduction to combat wasteful consumption (12.c). Table 3.4 provides details on the UN targets and indicators for goal 12, including the multipurpose indicators.

Table 3.2 UN targets and indicators for goal 9.

Targets	Indicators
9.1 Develop quality, reliable, sustainable, and resilient infrastructure, including regional and transborder infrastructure, to support economic development and human well-being, with a focus on affordable and equitable access for all.	9.1.1 Proportion of the rural population who live within 2 km of an all-season road. 9.1.2 Passenger and freight volumes, by mode of transport.
9.2 Promote inclusive and sustainable industrialization and, by 2030, significantly raise industry's share of employment and gross domestic product, in line with national circumstances, and double its share in least developed countries.	9.2.1 Manufacturing value added as a proportion of GDP and per capita. 9.2.2 Manufacturing employment as a proportion of total employment.
9.3 Increase the access of small-scale industrial and other enterprises, in particular in developing countries, to financial services, including affordable credit, and their integration into value chains and markets.	9.3.1 Proportion of small-scale industries in total industry value added. 9.3.2 Proportion of small-scale industries with a loan or line of credit.
9.4 By 2030, upgrade infrastructure and retrofit industries to make them sustainable, with increased resource-use efficiency and greater adoption of clean and environmentally sound technologies and industrial processes, with all countries taking action in accordance with their respective capabilities.	9.4.1 CO_2 emission per unit of value added.
9.5 Enhance scientific research and upgrade the technological capabilities of industrial sectors in all countries, in particular developing countries, including, by 2030, encouraging innovation and substantially increasing the number of research and development workers per 1 million people and public and private research and development spending.	9.5.1 Research and development expenditure as a proportion of GDP. 9.5.2 Researchers (in full-time equivalent) per million inhabitants.
9.a Facilitate sustainable and resilient infrastructure development in developing countries through enhanced financial, technological, and technical support to African countries, least developed countries, landlocked developing countries, and small island developing States.	9.a.1 Total official international support (official development assistance plus other official flows) to infrastructure.

(Continued)

Table 3.2 UN targets and indicators for goal 9.—cont'd

Targets	Indicators
9.b Support domestic technology development, research, and innovation in developing countries, including by ensuring a conducive policy environment for, inter alia, industrial diversification and value addition to commodities.	9.b.1 Proportion of medium and high-tech industry value added in total value added.
9.c Significantly increase access to information and communications technology and strive to provide universal and affordable access to the Internet in least developed countries.	9.c.1 Proportion of population covered by a mobile network, by technology.

EU indicators for goal 9

09_10 Gross domestic expenditure on R&D by sector.
09_30 R&D personnel by sector.
09_40 Patent applications to the European Patent Office (EPO).
09_50 Share of busses and trains in total passenger transport.
09_60 Share of rail and inland waterways in total freight transport.
09_70 Air emission intensity from industry.

Multi-purpose indicators

12_30 Average CO_2 emissions per km from new passenger cars.
17_60 High-speed internet coverage, by type of area

From SDG 09, KnowSDGs Platform, Retrieved from: https://knowsdgs.jrc.ec.europa.eu/sdg/9.

Table 3.3 UN targets and indicators for goal 11.

Targets	Indicators
11.1 By 2030, ensure access for all to adequate, safe, and affordable housing and basic services and upgrade slums.	11.1.1 Proportion of urban population living in slums, informal settlements, or inadequate housing.
11.2 By 2030, provide access to safe, affordable, accessible, and sustainable transport systems for all, improving road safety, notably by expanding public transport, with special attention to the needs of those in vulnerable situations, women, children, persons with disabilities, and older persons.	11.2.1 Proportion of population that has convenient access to public transport, by sex, age, and persons with disabilities.
11.3 By 2030, enhance inclusive and sustainable urbanization and capacity for participatory, integrated, and sustainable human settlement planning and management in all countries.	11.3.1 Ratio of land consumption rate to population growth rate. 11.3.2 Proportion of cities with a direct participation structure of civil society in urban planning and management that operate regularly and democratically.

Table 3.3 UN targets and indicators for goal 11.—cont'd

Targets	Indicators
11.4 Strengthen efforts to protect and safeguard the world's cultural and natural heritage.	11.4.1 Total per capita expenditure on the preservation, protection, and conservation of all cultural and natural heritage, by source of funding (public, private), type of heritage (cultural, natural), and level of government (national, regional, and local/municipal).
11.5 By 2030, significantly reduce the number of deaths and the number of people affected and substantially decrease the direct economic losses relative to global gross domestic product caused by disasters, including water-related disasters, with a focus on protecting the poor and people in vulnerable situations.	11.5.1 Number of deaths, missing persons, and persons affected by disaster per 100,000 people. 11.5.2 Direct economic loss in relation to global GDP, damage to critical infrastructure, and number of disruptions to basic services, attributed to disasters.
11.6 By 2030, reduce the adverse per capita environmental impact of cities, including by paying special attention to air quality and municipal and other waste management.	11.6.1 Proportion of municipal solid waste collected and managed in controlled facilities out of total municipal waste generated, by cities. 11.6.2 Annual mean levels of fine particulate matter (e.g., PM2.5 and PM10) in cities (population weighted).
11.7 By 2030, provide universal access to safe, inclusive, and accessible, green and public spaces, in particular for women and children, older persons, and persons with disabilities.	11.7.1 Average share of the built-up area of cities that is open space for public use for all, by sex, age, and persons with disabilities. 11.7.2 Proportion of persons victim of physical or sexual harassment, by sex, age, disability status, and place of occurrence, in the previous 12 months.
11.a Support positive economic, social, and environmental links between urban, peri-urban, and rural areas by strengthening national and regional development planning.	11.a.1 Number of countries that have national urban policies or regional development plans that (a) respond to population dynamics; (b) ensure balanced territorial development; and (c) increase local fiscal space.

(Continued)

Table 3.3 UN targets and indicators for goal 11.—cont'd

Targets	Indicators
11.b By 2020, substantially increase the number of cities and human settlements adopting and implementing integrated policies and plans toward inclusion, resource efficiency, mitigation and adaptation to climate change, resilience to disasters, and develop and implement, in line with the Sendai Framework for Disaster Risk Reduction 2015—2030, holistic disaster risk management at all levels.	11.b.1 Number of countries that adopt and implement national disaster risk reduction strategies in line with the Sendai Framework for Disaster Risk Reduction 2015—2030. 11.b.2 Proportion of local governments that adopt and implement local disaster risk reduction strategies in line with national disaster risk reduction strategies.
11.c Support least developed countries, including through financial and technical assistance, in building sustainable and resilient buildings utilizing local materials.	11.c.1 Proportion of financial support to the least developed countries that is allocated to the construction and retrofitting of sustainable, resilient, and resource-efficient buildings utilizing local materials.

EU indicators for goal 11

11_10 Overcrowding rate by poverty status.
11_20 Population living in households considering that they suffer from noise, by poverty status.
11_31 Settlement area per capita.
11_40 Road traffic deaths, by type of roads.
11_50 Exposure to air pollution by particulate matter.
11_60 Recycling rate of municipal waste.

Multi-purpose indicators

01_60 Population living in a dwelling with a leaking roof, damp walls, floors, or foundation or rot in window frames of the floor by poverty status.
06_20 Population connected to at least secondary wastewater treatment.
09_50 Share of busses and trains in total passenger transport.
16_20 Population reporting occurrence of crime, violence, or vandalism in their area by poverty status.

From SDG 11, KnowSDGs Platform, Retrieved from: https://knowsdgs.jrc.ec.europa.eu/sdg/11.

Table 3.4 UN targets and indicators for goal 12.

Targets	Indicators
12.1 Implement the 10-year framework of programs on sustainable consumption and production, all countries taking action, with developed countries taking the lead, taking into account the development and capabilities of developing countries.	12.1.1 Number of countries developing, adopting, or implementing policy instruments aimed at supporting the shift to sustainable consumption and production.
12.2 By 2030, achieve the sustainable management and efficient use of natural resources.	12.2.1 Material footprint, material footprint per capita, and material footprint per GDP. 12.2.2 Domestic material consumption, domestic material consumption per capita, and domestic material consumption per GDP.
12.3 By 2030, halve per capita global food waste at the retail and consumer levels and reduce food losses along production and supply chains, including postharvest losses.	12.3.1 (a) Food loss index and (b) food waste index.
12.4 By 2020, achieve the environmentally sound management of chemicals and all wastes throughout their life cycle, in accordance with agreed international frameworks, and significantly reduce their release to air, water, and soil to minimize their adverse impacts on human health and the environment.	12.4.1 Number of parties to international multilateral environmental agreements on hazardous waste, and other chemicals that meet their commitments and obligations in transmitting information as required by each relevant agreement. 12.4.2 (a) Hazardous waste generated per capita; and (b) proportion of hazardous waste treated, by type of treatment
12.5 By 2030, substantially reduce waste generation through prevention, reduction, recycling, and reuse.	12.5.1 National recycling rate, tons of material recycled.
12.6 Encourage companies, especially large and transnational companies, to adopt sustainable practices and to integrate sustainability information into their reporting cycle.	12.6.1 Number of companies publishing sustainability reports.
12.7 Promote public procurement practices that are sustainable, in accordance with national policies and priorities.	12.7.1 Degree of sustainable public procurement policies and action plan implementation.

(Continued)

Table 3.4 UN targets and indicators for goal 12.—cont'd

Targets	Indicators
12.8 By 2030, ensure that people everywhere have the relevant information and awareness for sustainable development and lifestyles in harmony with nature.	12.8.1 Extent to which (i) global citizenship education and (ii) education for sustainable development (including climate change education) are mainstreamed in (a) national education policies; (b) curricula; (c) teacher education; and (d) student assessment.
12.a Support developing countries to strengthen their scientific and technological capacity to move toward more sustainable patterns of consumption and production.	12.a.1 Installed renewable energy-generating capacity in developing countries (in watts per capita).
12.b Develop and implement tools to monitor sustainable development impacts for sustainable tourism that creates jobs and promotes local culture and products.	12.b.1 Implementation of standard accounting tools to monitor the economic and environmental aspects of tourism sustainability.
12.c Rationalize inefficient fossil-fuel subsidies that encourage wasteful consumption by removing market distortions, in accordance with national circumstances, including by restructuring taxation and phasing out those harmful subsidies, where they exist, to reflect their environmental impacts, taking fully into account the specific needs and conditions of developing countries and minimizing the possible adverse impacts on their development in a manner that protects the poor and the affected communities.	12.c.1 Amount of fossil-fuel subsidies per unit of GDP (production and consumption).

EU indicators for goal 12

12_10 Consumption of chemicals by hazardousness—EU aggregate.
12_20 Resource productivity and domestic material consumption (DMC).
12_30 Average CO_2 emissions per km from new passenger cars.
12_41 Circular material use rate.
12_50 Generation of waste excluding major mineral wastes by hazardousness.
12_61 Gross value added in environmental goods and services sector.

Multi-purpose indicators

07_30 Energy productivity.

From SDG 12 KnowSDGs Platform, Retrieved from: https://knowsdgs.jrc.ec.europa.eu/sdg/12.

Goal[a]	Code[b]	MPI[c]	Indicator name[d]	Unit(s)[e]	Frequency of data collection[f]	Geographical coverage[g]	Data source[h]	Data provider[i]
1	**Goal 1. End poverty in all its forms everywhere**							
1	01_10		People at risk of poverty or social exclusion	% of population and thousand persons i.Total ii.Less than 18 years	Every year	EU aggregate and all MS; plus other countries	ESS (SILC)	Eurostat
1	01_20		People at risk of monetary poverty after social transfers	% of population and thousand persons	Every year	EU aggregate and all MS; plus other countries	ESS (SILC) and ECHP	Eurostat
1	01_31		Severe material and social deprivation rate (SMSD)	% of population and thousand persons i.Total ii.Less than 18 years	Every year	EU aggregate and all MS; plus other countries	ESS (SILC)	Eurostat
1	01_40		People living in households with very low work intensity	% of population and thousand persons i.Age group less than 65 ii.Age group less than 18	Every year	EU aggregate and all MS; plus other countries	ESS (SILC)	Eurostat
1	01_41	mpi -> 8	In work at-risk-of-poverty rate	% of employed persons aged 18 or over	Every year	EU aggregate and all MS; plus other countries	ESS (SILC)	Eurostat
1	01_50		Housing cost overburden rate	% Of population i.Total ii.Below 60% of median equivalised income iii.Above 60% of median equivalised income	Every year	EU aggregate and all MS; plus other countries	ESS (SILC)	Eurostat

(Continued)

List of indicators for 2023 monitoring report.—cont'd

Goal[a]	Code[b]	MPI[c]	Indicator name[d]	Unit(s)[e]	Frequency of data collection[f]	Geographical coverage[g]	Data source[h]	Data provider[i]
2			**Goal 2. End hunger, achieve food security and improved nutrition, and promote sustainable agriculture**					
2	02_10	mpi -> 3	Obesity rate	% of population aged 18 or over i.Overweight (BMI >25) ii.Preobese (BMI 25–30) iii.Obese (BMI>30)	More than 3 years	EU aggregate and all MS; plus other countries	ESS (EHIS and SILC)	Eurostat
2	02_20		Agricultural factor income per annual work unit (AWU)	Index 2010 = 100 and chain-linked volumes (2010) in EUR	Every year	EU aggregate and all MS; plus other countries	ESS (EAA)	Eurostat, DG AGRI
2	02_30		Government support to agricultural research and development	Million EUR and EUR per inhabitant (current prices)	Every year	EU aggregate and all MS; plus other countries	ESS (GBARD) Government budget allocations for R&D	Eurostat; OECD
2	02_40		Area under organic farming	% of total utilized agricultural area (UAA)	Every year	EU aggregate and all MS; plus other countries	ESS	Eurostat
2	02_60		Ammonia emissions from agriculture	tonnes and kg per hectare utilized agricultural area (UAA)	Every year	EU aggregate and all MS; plus other countries	Reporting under National Emission Ceilings Directive (NECD) and Convention on Long-range Transboundary Air Pollution (CLRTAP)	EEA

Goal 3. Ensure healthy lives and promote well-being for all at all ages

3							
3	03_11	Healthy life years at birth	Years i. Total ii. Males iii. Females	Every year	EU aggregate and all MS; plus other countries	ESS (SILC)	Eurostat
3	03_20	Share of people with good or very good perceived health	% of population aged 16 or over i. Total ii. Males iii. Females	Every year	EU aggregate and all MS; plus other countries	ESS (SILC)	Eurostat
3	03_30	Smoking prevalence	% of population aged 15 or over i. Total ii. Males iii. Females	Every 3 years	EU aggregate and all MS	Eurobarometer	DG SANTE
3	03_41	Standardized death rate due to tuberculosis, HIV, and hepatitis	Number per 100,000 persons i. Total ii. Tuberculosis iii. Hepatitis iv. HIV	Every year	EU aggregate and all MS; plus other countries	ESS	Eurostat
3	03_42	Standardized avoidable mortality	Number per 100,000 persons aged less than 75 years i. Total ii. Preventable mortality iii. Treatable mortality	Every year	EU aggregate and all MS; plus other countries	ESS	Eurostat

(Continued)

List of indicators for 2023 monitoring report.—cont'd

Goal[a]	Code[b]	MPI[c]	Indicator name[d]	Unit(s)[e]	Frequency of data collection[f]	Geographical coverage[g]	Data source[h]	Data provider[i]
3	03_60	mpi -> 1	Self-reported unmet need for medical care	% of population aged 16 and over i.Total ii.Males iii.Females	Every year	EU aggregate and all MS; plus other countries	ESS (SILC)	Eurostat
4			**Goal 4. Ensure inclusive and equitable quality education and promote lifelong learning opportunities for all**					
4	04_40		Low achievement in reading, maths, and science	% of 15-year-old students i.Reading ii.Maths iii.Science	More than 3 years	EU aggregate and all MS; plus other countries	PISA	OECD
4	04_30		Participation in early childhood education by sex	% of the age group between 3 years old and the starting age of compulsory primary education i.Total ii.Males iii.Females	Every year	EU aggregate and all MS; plus other countries	ESS (UOE)	Eurostat
4	04_10	mpi -> 5	Early leavers from education and training	% of population aged 18 –24 i.Total ii.Males iii.Females	Every year	EU aggregate and all MS; plus other countries	ESS (LFS)	Eurostat

4	04_20 mpi -> 5; 9	Tertiary educational attainment Y25–34	% of population aged 25 to 34	Every year	EU aggregate and all MS; plus other countries	ESS (LFS)	Eurostat
4	04_60	Adult participation in learning	% of population aged 25 to 64 i.Total ii.Males iii.Females	Every year	EU aggregate and all MS; plus other countries	ESS (LFS)	Eurostat
4	04_70	Share of adults having at least basic digital skills	% of individuals aged 16–74	Every 2 years	EU aggregate and all MS; plus other countries	ESS	Eurostat
5	**Goal 5. Achieve gender equality and empower all women and girls**						
5	05_10	Physical and sexual violence to women	% of women i.Age 15–74 ii.Age 18–29 iii.Age 30–39 iv.Age 40–49 v.Age 50–59 vi.Age 60+	a-periodic	EU aggregate and all MS	Survey on violence against women http://fra.europa.eu/en/publications-and-resources/data-and-maps/survey-data-explorer-violence-against-women-survey	DG JUST; EU agency for fundamental rights
5	05_20	Gender pay gap in unadjusted form	% of average gross hourly earnings of men	Every year	EU aggregate and most MS; plus other countries	ESS (SES)	Eurostat

(Continued)

List of indicators for 2023 monitoring report.—cont'd

Goal[a]	Code[b]	MPI[c]	Indicator name[d]	Unit(s)[e]	Frequency of data collection[f]	Geographical coverage[g]	Data source[h]	Data provider[i]
5	05_30		Gender employment gap	Percentage points i.Employed persons ii.Employed persons working part time iii.Employed persons with a temporary contract iv.Underemployed persons working part time	Every year	EU aggregate and all MS; plus other countries	ESS (LFS)	Eurostat
5	05_40	mpi -> 8	People outside the labor force due to caring responsibilities	% of population aged 20 to 64 i.Total ii.Males iii.Females	Every year	EU aggregate and all MS; plus other countries	ESS (LFS)	Eurostat
5	05_50		Seats held by women in national parliaments and governments	% of seats i.National parliaments ii.National governments	Every year	EU aggregate and all MS; plus other countries	The Gender Statistics Database (GSD)	EIGE
5	05_60		Positions held by women in senior management	% of positions i.Board members ii.Executives	Every year	EU aggregate and all MS; plus other countries	The Gender Statistics Database (GSD)	EIGE

Goal 6. Ensure availability and sustainable management of water and sanitation for all

6	06_10	Population having neither a bath, nor a shower, nor indoor flushing toilet in their household	% of population i.Total ii.Below 60% of median equivalised income iii.Above 60% of median equivalised income	Every 3 years	EU aggregate and all EU MS; plus other countries	ESS (SILC)	Eurostat
6	06_20 mpi -> 11	Population connected to at least secondary wastewater treatment	% of population	Every year	EU aggregate and most MS; plus other countries	OECD/ESTAT joint questionnaire	Eurostat
6	06_30 mpi -> 15	Biochemical oxygen demand in rivers	mg O_2 per liter	Every year	<75% of MS or no EU aggregate	Waterbase – water quality	EEA
6	06_40 mpi -> 2	Nitrate in groundwater	mg NO_3 per liter	Every year	EU aggregate and some MS; plus other countries	Waterbase – water quality	EEA
6	06_50 mpi -> 15	Phosphate in rivers	mg PO_4 per liter	Every year	EU aggregate and <75% MS; plus other countries	Waterbase – water quality	EEA
6	06_60	Water exploitation index, plus (WEI+)	% of long-term average available water (LTAA)	Every 2 years	EU aggregate and all MS; plus other countries	Waterbase –water quality	EEA, data collected by Eurostat

(Continued)

List of indicators for 2023 monitoring report.—cont'd

Goal[a]	Code[b]	MPI[c]	Indicator name[d]	Unit(s)[e]	Frequency of data collection[f]	Geographical coverage[g]	Data source[h]	Data provider[i]
7	**Goal 7. Ensure access to affordable, reliable, sustainable, and modern energy for all**							
7	07_10		Primary and final energy consumption	Million tonnes of oil equivalent index 2005 = 100 and tonnes of oil equivalent per capita	Every year	EU aggregate and all MS; plus other countries	ESS	Eurostat
7	07_20		Final energy consumption in households per capita	Kg of oil equivalent	Every year	EU aggregate and all MS; plus other countries	ESS	Eurostat
7	07_30	mpi -> 12	Energy productivity	Chain-linked volumes (2010) in EUR and PPS per kg of oil equivalent	Every year	EU aggregate and all MS; plus other countries	ESS	Eurostat
7	07_40	mpi -> 13	Share of renewable energy in gross final energy consumption	% i.All sectors ii.Transport iii.Electricity iv.Heating and cooling	Every year	EU aggregate and all MS; plus other countries	ESS (SHARES)	Eurostat
7	07_50		Energy import dependency	% of imports in total gross available energy i.All products ii.Solid fossil fuels iii.Total petroleum products iv.Natural gas	Every year	EU aggregate and all MS; plus other countries	ESS	Eurostat

7	07_60	Population unable to keep home adequately warm	% of population i.Total ii.Below 60% of median equivalised income iii.Above 60% of median equivalised income	EU aggregate and all MS; plus other countries	Every year	ESS (SILC)	Eurostat

Goal 8. Promote sustained, inclusive, and sustainable economic growth, full and productive employment, and decent work for all

8	08_10	Real GDP per capita	Chain-linked volumes (2010) in EUR and % change on the previous year	EU aggregate and all MS; plus other countries	Every year	ESS (National accounts)	Eurostat
8	08_11	Investment share of GDP	% of GDP i.Total investment ii.Business investment iii.Government investment iv.Households investments	EU aggregate and all MS; plus other countries	Every year	ESS	Eurostat
8	08_20	Young people neither in employment nor in education and training	% of population aged 15–29 i.Total ii.Males iii.Females	EU aggregate and all MS; plus other countries	Every year	ESS (LFS)	Eurostat

(Continued)

List of indicators for 2023 monitoring report.—cont'd

Goal[a]	Code[b]	MPI[c]	Indicator name[d]	Unit(s)[e]	Frequency of data collection[f]	Geographical coverage[g]	Data source[h]	Data provider[i]
8	08_30		Employment rate	% of population aged 20–64 i. Total ii. Males iii. Females	Every year	EU aggregate and all MS; plus other countries	ESS (LFS)	Eurostat
8	08_40		Long-term unemployment rate	% of population in the labor force i. Total ii. Males iii. Females	Every year	EU aggregate and all MS; plus other countries	ESS (LFS)	Eurostat
8	08_60	mpi -> 3	Fatal accidents at work per 100,000 workers	Number per 100,000 employees i. Total ii. Males iii. Females	Every year	EU aggregate and all MS; plus other countries	ESS (ESAW)	Eurostat
9			**Goal 9. Build resilient infrastructure, promote inclusive and sustainable industrialization, and foster innovation**					
9	09_10		Gross domestic expenditure on R&D	% of GDP i. Total ii. Business enterprise sector iii. Government sector iv. Higher education sector v. Private nonprofit sector	Every year	EU aggregate and all MS; plus other countries	ESS	Eurostat

9	09_30	R&D personnel	% of population in the labor force i.Total ii.Business enterprise sector iii.Government sector iv.Higher education sector v.Private nonprofit sector	Every year	EU aggregate and all MS; plus other countries	OECD/ESTAT joint questionnaire	Eurostat
9	09_40	Patent applications to the European Patent Office (EPO)	Total number and number per million inhabitants i.Applicants ii.Inventors	Every year	EU aggregate and all MS; plus other countries	EPO annual reports	EPO
9	09_50 mpi -> 11	Share of buses and trains in inland passenger transport	% of passenger-kilometres i.All collective transport modes ii.Trains iii.Motor coaches, buses, and trolley buses	Every year	EU aggregate and all MS; plus other countries	ESS	Eurostat
9	09_60	Share of rail and inland waterways in inland freight transport	% of tonne-kilometres i.All railways and inland waterways ii.Railways iii.Inland waterways	Every year	EU aggregate and all MS; plus other countries	ESS	Eurostat

(Continued)

List of indicators for 2023 monitoring report.—cont'd

Goal[a]	Code[b]	MPI[c]	Indicator name[d]	Unit(s)[e]	Frequency of data collection[f]	Geographical coverage[g]	Data source[h]	Data provider[i]
9	09_70		Air emission intensity from industry	Grams per euro, chain-linked volumes (2010) i.Particulates <2.5 μm ii.Particulates <10 μm	Every year	EU aggregate and all MS and other countries	ESS; Air Emission Accounts (AEA)	Eurostat
10	**Goal 10. Reduce inequality within and among countries**							
10	10_10		Disparities in GDP per capita	PPS (current prices), index EU27 = 100 and coefficient of variation	Every year	EU aggregate and all MS; plus other countries	ESS (National accounts)	Eurostat
10	10_20		Disparities in household income per capita	PPS (current prices) and index EU27 = 100	Every year	EU aggregate and most MS; plus other countries	ESS	Eurostat
10	10_30 mpi -> 1		Relative median at-risk-of-poverty gap	% distance to poverty threshold	Every year	EU aggregate and all MS; plus other countries	ESS (SILC)	Eurostat
10	10_41		Income distribution - income quintile share ratio	Quintile share ratio	Every year	EU aggregate and all MS; plus other countries	ESS (SILC)	Eurostat
10	10_50		Income share of the bottom 40% of the population	% of income	Every year	EU aggregate and all MS; plus other countries	ESS (SILC)	Eurostat
10	10_60		Asylum applications	Number per million inhabitants i.First time application ii.Positive first instance decision	Every year	EU aggregate and all MS; plus other countries	ESS	Eurostat

Goal 11. Make cities and human settlements inclusive, safe, resilient, and sustainable

11						
11	11_11 mpi -> 1	Severe housing deprivation rate	% of population i.Total ii.Below 60% of median equivalised income iii.Above 60% of median equivalised income	Every 3 years	EU aggregate and all MS; plus other countries	ESS (SILC)	Eurostat
11	11_20 mpi -> 3	Population living in households considering that they suffer from noise	% of population i.Total ii.Below 60% of median equivalised income iii.Above 60% of median equivalised income	Every 3 years	EU aggregate and all MS; plus other countries	ESS (SILC)	Eurostat
11	11_31	Settlement area per capita	Square meters per capita	More than 3 years	EU aggregate and all MS	ESS (LUCAS)	Eurostat
11	11_40 mpi -> 3	Road traffic deaths	Persons and number per 100,000 persons i.Total ii.Motorways iii.Urban roads iv.Rural roads v.Unknown	Every year	EU aggregate and all MS; plus other countries	CARE database	DG MOVE

(Continued)

List of indicators for 2023 monitoring report.—cont'd

Goal[a]	Code[b]	MPI[c]	Indicator name[d]	Unit(s)[e]	Frequency of data collection[f]	Geographical coverage[g]	Data source[h]	Data provider[i]
11	11_52	mpi -> 3	Premature deaths due to exposure to fine particulate matter (PM2.5)	Number and number per 100,000 people (rate)	Every year	EU aggregate and all MS; plus other countries	DG ENV	EEA
11	11_60		Recycling rate of municipal waste	% of total municipal waste generated	Every year	EU aggregate and all MS; plus other countries	ESS	Eurostat
12	**Goal 12. Ensure sustainable consumption and production patterns**							
12	12_10		New name: "Consumption of hazardous chemicals" Former name was "Consumption of hazardous and nonhazardous chemicals"	Million tonnes i.Hazardous and nonhazardous—total ii.Hazardous total iii.Hazardous to health iv.Hazardous to environment	Every year	Only EU aggregate; no MS data available.	ESS (PRODCOM; COMEXT)	Eurostat
12	12_21	mpi -> 8	Material footprint	Thousand tonnes and tonnes per capita	Every year	EU aggregate and all MS	ESS	Eurostat
12	12_30	mpi -> 13	Average CO_2 emissions per km from new passenger cars	g CO_2 per km	Every year	EU aggregate and all MS	Reporting under Regulation (EC) No 443/2009	EEA/DG CLIMA

12	12_41	Circular material use rate	% of total material use	Every year	EU aggregate and all MS	ESS	Eurostat
12	12_51	Generation of waste	kg per capita i.Hazardous and nonhazardous—total ii.Hazardous iii.Nonhazardous	Every 2 years	EU aggregate and all MS; plus other countries	ESS	Eurostat
12	12_61 mpi -> 9	Gross value added in the environmental goods and services sector	Chain-linked volumes (2010) in EUR and % of GDP	Every year	EU aggregate and most MS; plus other countries	ESS	Eurostat
13	**Goal 13. Take urgent action to combat climate change and its impacts**						
13	13_10	Net greenhouse gas emissions	Index 1990 = 100 and tonnes of CO_2 equivalent per capita	Every year	EU aggregate and all MS; plus other countries	UNFCCC reporting	EEA
13	13_21	Net greenhouse gas emission of the Land Use, Land Use Change, and Forestry (LULUCF) sector	Thousand tonnes of CO_2 equivalent; tonnes of CO_2 equivalent per capita; tonnes of CO_2 equivalent per km^2	Every year	EU aggregate and all MS; plus other countries	UNFCCC reporting	EEA

(Continued)

List of indicators for 2023 monitoring report.—cont'd

Goal[a]	Code[b]	MPI[c]	Indicator name[d]	Unit(s)[e]	Frequency of data collection[f]	Geographical coverage[g]	Data source[h]	Data provider[i]
13	13_40		Climate-related economic losses	Million EUR (current prices) and EUR per capita i.Annual values ii.Average losses over 30 years	Every year	EU aggregate and all MS	CATDAT - Risklayer	EEA
13	13_50		Contribution to the international 100bn USD commitment on climate-related expenses	Million EUR (current prices)	Every year	EU aggregate and all MS; plus EIB and European Commission	Regulation (EU) 2018/1999 on the Governance of the Energy Union and Climate Action	DG CLIMA; Eionet
13	13_60		Population covered by the Covenant of Mayors for Climate and Energy signatories	Million persons and % of population	Every year	EU aggregate and all MS; plus other countries	The Covenant of Mayors for Climate & Energy	JRC
14	**Goal 14. Conserve and sustainably use the oceans, seas, and marine resources for sustainable development**							
14	14_10		Marine protected areas	km^2 and %	a-periodic	EU aggregate and all MS (except landlocked)	ETC/BD	EEA

14	14_21	Estimated trends in fish stock biomass	Fish stock biomass—index 2003 = 100 i.EU total marine waters ii.North East Atlantic iii.Mediterranean and Black Sea	Every year	EU aggregate (EU total marine waters); for MS not applicable (only FAO fishing areas NE Atlantic, Mediterranean, and Black Sea)	STECF	JRC
14	14_30	Estimated trends in fishing pressure	Fish stocks—model-based median value i.EU total marine waters ii.North East Atlantic iii.Mediterranean and Black Sea	Every year	EU aggregate (EU total marine waters); for MS not applicable (only FAO fishing areas NE Atlantic, Mediterranean, and Black Sea)	STECF	JRC
14	14_40 mpi -> 6	Bathing sites with excellent water quality	Number and % of bathing sites i.Coastal water ii.Inland water	Every year	EU aggregate and all MS (coastal water: except landlocked countries)	EEA	EEA
14	14_50	Global mean surface seawater acidity	pH value	Every year	Not applicable	CMEMS Copernicus Marine Service	Mercator Ocean International (Moi)
14	14_60	Marine waters affected by eutrophication	km^2 % of the Exclusive Economic Zones (EEZ)	Every year	EU aggregate and per MS (except landlocked countries)	CMEMS Copernicus Marine Service	Mercator Ocean International (Moi)

(Continued)

List of indicators for 2023 monitoring report.—cont'd

Goal[a]	Code[b]	MPI[c]	Indicator name[d]	Unit(s)[e]	Frequency of data collection[f]	Geographical coverage[g]	Data source[h]	Data provider[i]
15			**Goal 15. Protect, restore, and promote sustainable use of terrestrial ecosystems, sustainably manage forests, combat desertification, and halt and reverse land degradation and halt and reverse biodiversity loss**					
15	15_10		Share of forest area	% of total area i. All forest area FAO ii. Forest FAO iii. Other wooded land FAO	More than 3 years	EU aggregate and all MS	ESS (LUCAS)	Eurostat
15	15_20		Terrestrial protected areas	km² and %	Every year	EU aggregate and all MS	DG ENV	EEA
15	15_41		Soil sealing index	Index 2006 = 100; % of total surface; km² of sealed surface	Every 3 years	EU aggregate and all MS	Copernicus HRL	EEA
15	15_50 mpi -> 2		Estimated soil erosion by water - area affected by severe erosion rate	km² and % of potential erodible area	a-periodic	EU aggregate and all MS	Soil erosion database	JRC
15	15_60 mpi -> 2		Common bird index	Index 2000 = 100 and index 1990 = 100 i. All common species ii. Common farmland species iii. Common forest species	Every year	Only EU aggregate; no MS data available.	EBCC/RSPB/BirdLife/Czech Society for ornithology	European Bird Census Council

15	15_61	Grassland butterfly index	Index 2000 = 100 and index 1991 = 100	Every year	Only EU aggregate; no MS data available.	BMS (Butterfly Monitoring Scheme)	EEA (Butterfly Conservation Europe)
16	**Goal 16. Promote peaceful and inclusive societies for sustainable development, provide access to justice for all, and build effective, accountable, and inclusive institutions at all levels**						
16	16_10	Standardized death rate due to homicide	Number per 100,000 persons i.Total ii.Males iii.Females	Every year	EU aggregate and all MS; plus other countries	ESS	Eurostat
16	16_20 mpi -> 11	Population reporting occurrence of crime, violence or vandalism in their area	% of population i.Total ii.Below 60% of median equivalised income iii.Above 60% of median equivalised income	Every 3 years	EU aggregate and all MS; plus other countries	ESS (SILC)	Eurostat
16	16_30	General government total expenditure on law courts	Million EUR and EUR per capita (current prices)	Every year	EU aggregate and all MS; plus other countries	ESS	Eurostat

(Continued)

List of indicators for 2023 monitoring report.—cont'd

Goal[a]	Code[b]	MPI[c]	Indicator name[d]	Unit(s)[e]	Frequency of data collection[f]	Geographical coverage[g]	Data source[h]	Data provider[i]
16	16_40		Perceived independence of the justice system	% of population i. Very good or fairly good ii. Very good iii. Fairly good iv. Very bad or fairly bad v. Very bad vi. Fairly bad vii. Unknown	Every year	EU aggregate and all MS	Eurobarometer	DG COMM
16	16_50		Corruption Perceptions Index	Score scale of 0 (highly corrupt) to 100 (very clean)	Every year	EU aggregate and all MS; plus other countries	http://www.transparency.org/research/cpi/overview	Transparency International
16	16_60		Population with confidence in EU institutions	% of population i. European Parliament ii. European Commission iii. European Central Bank	Every year	EU aggregate and all MS; plus other countries	Eurobarometer	DG COMM
17			**Goal 17. Strengthen the means of implementation and revitalize the Global Partnership for Sustainable Development**					
17	17_10		Official development assistance as a share of gross national income	% of GNI (at current prices) i. DAC countries ii. Least developed countries	Every year	EU aggregate and all MS; plus other countries	OECD database	DG INTPA; OECD (DAC)

17	17_20	EU financing to developing countries	Million EUR (current prices) i.Official development assistance ii.Grants by NGOs iii.Private flows iv.Other official flows v.Officially supported export credits	Every year	EU aggregate and all MS; plus other countries	OECD database	OECD (DAC)
17	17_30	EU imports from developing countries	Million EUR (current prices) i.DAC countries ii.Least developed countries iii.Lower middle income countries iv.Other low income countries v.Upper middle income countries excl. China vi.China (excl. Hong Kong)	Every year	EU aggregate and all MS	ESS	Eurostat
17	17_40	General government gross debt	% of GDP and million EUR (current prices)	Every year	EU aggregate and all MS	ESS	Eurostat

(Continued)

List of indicators for 2023 monitoring report.—cont'd

Goal[a]	Code[b]	MPI[c]	Indicator name[d]	Unit(s)[e]	Frequency of data collection[f]	Geographical coverage[g]	Data source[h]	Data provider[i]
17	17_50		Shares of environmental taxes in total tax revenues	% (current prices)	Every year	EU aggregate and all MS; plus other countries	ESS	Eurostat
17	17_60	mpi -> 9	Share of households with high-speed internet connection	% Of households i.Total ii.low settled areas	Every year	EU aggregate and all MS, and other countries	IHS Markit, Omdia, Point Topic and VVA, Broadband coverage in Europe studies	DG CNECT

[a]Can be used to select each of the 17 goals
[b]Indicator code in Eurostat's database
[c]Multipurpose indicator indicates the goal(s) to which the indicator is also allocated for monitoring purposes
[d]Name of indicator
[E]Unit(s) of measurements as presented in Eurostat's database
[f]Categories are: every year; every 2 years; every 3 years; every >3 years; a-periodic
[g]Geographical coverage
[h]Data collection, survey, etc. on which the data for the indicator is based
[i]Service or institution that disseminates the indicator or the underlying data

References

Eurostat Unit. (2023). *EU SDG Indicator set 2023. Result of the review in preparation of the 2023 edition of the EU SDG monitoring report.* Retrieved from: https://ec.europa.eu/eurostat/documents/276524/15877045/EU-SDG-indicators-2023.pdf/31e0dfe2-253b-73b7-fb72-408c7208cb85?t=1673438075710.

Eurostat. (2023). *Sustainable development in the European Union Monitoring report on progress towards the SDGs in an EU context* (2023 edition). Flagship publications, ISBN 978-92-68-00374-9 https://doi.org/10.2785/403194. Retrieved from: https://ec.europa.eu/eurostat/web/products-flagship-publications/w/ks-04-23-184.

ISTAT. (2022). *SDGs report. Statistical information for 2030 Agenda in Italy, Roma.* Retrieved from: https://www.istat.it/en/archivio/284043.

ISTAT, Rapporto SDGS. (2023). *Informazioni statistiche per l'agenda 2030 in Italia, Roma* (pp. 1–224). Retrieved from: https://www.istat.it/storage/rapporti-tematici/sdgs/2023/Rapporto-SDGs-2023.pdf.

KnowSDGs platform, Retrieved from: https://knowsdgs.jrc.ec.europa.eu/sdg/12.

Resolution adopted by the general assembly on 6 July 2017 [without reference to a main committee (A/71/L.75)] 71/313. Work of the statistical commission pertaining to the 2030 agenda for sustainable development. Retrieved from: https://ggim.un.org/documents/a_res_71_313.pdf.

SDG 09, KnowSDGs platform, Retrieved from: https://knowsdgs.jrc.ec.europa.eu/sdg/9.

SDG 11, KnowSDGs platform, Retrieved from: https://knowsdgs.jrc.ec.europa.eu/sdg/11.

SDG 12 KnowSDGs platform, Retrieved from: https://knowsdgs.jrc.ec.europa.eu/sdg/12.

Sustainable development goal indicators, Retrieved from: https://unstats.un.org/sdgs/iaeg-sdgs/.

UNCTAD. (2022). *Guidance on core indicators for sustainability and SDG impact reporting.* https://unctad.org/system/files/official-document/diae2022d1_en.pdf.

United Nations. (2023). *E-handbook on the sustainable development goals indicators – 150219.*

United Nations, The sustainable development goals report 2023, special edition. Retrieved from: https://unstats.un.org/sdgs/report/2023/.

CHAPTER FOUR

Sustainability reporting standards and guidelines

1. European sustainability reporting standards

The responsibility of firms in the prevention, management, and mitigation of social and environmental damage falls under the concept of corporate social responsibility (CSR) or responsible business conduct (Das et al., 2021; Fallah Shayan et al., 2022; Latapí Agudelo et al., 2019). The European Green Deal includes a series of financial measures to promote sustainable growth and pursue the objective of climate neutrality by 2050. To evaluate the achievement of these sustainability goals and direct financial and capital flows toward sustainable investments, firms should be able to produce, process, and report nonfinancial data and information (Hussain et al., 2020). These include information related to the environment, social issues, inclusivity and respect for human rights, transparency, and actions to combat corruption (Zhao et al., 2021).

Directive 2014/95/EU on nonfinancial information disclosure and diversity, commonly known as the "Non-Financial Reporting Directive"—NFRD, has established the groundwork for starting a path towards corporate responsibility and greater transparency of social and environmental information provided by firms across all sectors to achieve a high level of nonfinancial information disclosure.

The NFRD specifically targets large listed companies, banks, and insurance companies with more than 500 employees. Such public interest entities are required to publish reports and disclose information on corporate social responsibility policies, including due diligence processes, respect for human rights, fundamental freedoms, democratic principles as outlined in the United Nations International Charter of Human Rights, organizational employee well-being, gender diversity on boards of directors, risk management, and key performance indicators (KPIs). Reports on gender equality and equal pay issues should contain information on the gender pay gap, while those addressing employment and inclusion of people with disabilities issues should specify the accessibility measures taken by the firm.

Being a Sustainable Firm
ISBN: 978-0-443-14062-4
https://doi.org/10.1016/B978-0-443-14062-4.00004-0

The NFRD does not impose the use of a specific standard or framework for non-financial reporting, nor specific disclosure requirements. It recognizes that companies have high flexibility in choosing and implementing methods of disclosing information. This means that firms can disclose information they consider relevant and in the way they deem most appropriate to their context.

The Directive 2014/95/EU of the European Parliament and of the Council of October 22, 2014, amending Directive 2013/34/EU, provides specific indications on the information and documents constituting the Non-Financial Declaration. Specifically, article 1 states: "Large undertakings which are public-interest entities exceeding on their balance sheet dates the criterion of the average number of 500 employees during the financial year shall include in the management report a nonfinancial statement containing information to the extent necessary for an understanding of the undertaking's development, performance, position and impact of its activity, relating to, as a minimum, environmental, social and employee matters, respect for human rights, anticorruption and bribery matters, including.

(a) a brief description of the undertaking's business model;

(b) a description of the policies pursued by the undertaking in relation to those matters, including due diligence processes implemented;

(c) the outcome of those policies;

(d) the principal risks related to those matters linked to the undertaking's operations including, where relevant and proportionate, its business relationships, products or services which are likely to cause adverse impacts in those areas, and how the undertaking manages those risks;

(e) nonfinancial key performance indicators relevant to the particular business".

Recently, Directive 2022/2464/UE of the European Parliament and of the Council of December 14, 2022, amending Regulation (EU) No 537/2014, Directive 2004/109/EC, Directive 2006/43/EC, and Directive 2013/34/EU, introduced additional EU rights, rules, and regulations that sustainability reporting standards should take into account.

The sustainability reporting standards consider the reporting obligations established by Directive 2013/34/EU regarding the communication of information to users on sustainable development and company performance, highlighting the connections between various reported information. Additionally, they have to be consistent with the European law including the disclosure obligations established by regulation 2019/2088/EU, the indicators and methodologies established in regulations 2020/852/EU and 2016/

1011/EU, and the rules on minimum requirements for defining EU climate transition indices in implementation of Pillar III of Regulation 2013/575/EU. Sustainability reporting standards should also be consistent with Union environmental law (EC Regulation no. 1221/2009 and Directive 2003/87/EC of the European Parliament and of the Council, Recommendation 2013/179/EU of the Commission and subsequent amendments) and with other relevant Union rights, such as Directive 2010/75/EU of the European Parliament and of the Council.

To avoid penalizing firms subject to sustainability reporting obligations with additional administrative burdens and disruptions to their activities, it is essential to consider sustainability information, standards, and accounting, already adopted where appropriate, as sources of good practice. Such frameworks can include a wide range of models and procedures; among the most widespread are the International Accounting Standards Board, the Sustainability Accounting Standards Board, the Global Reporting Initiative, the International Integrated Reporting Council, the Carbon Disclosure Standards Board, and the Task Force on Climate-Related Financial Disclosures.

Sustainability reporting information is tipically presented in a specific section of the management report, covering eight key areas as described below.

1. Brief description of the firm's business model and strategy, highlighting the resilience of these aspects with respect to risks related to sustainability issues; opportunities linked to sustainability; implementation, financial, and investment plans; impact of the sustainability strategy on stakeholders' interests; and methods employed for implementing sustainability strategies.

2. Description of time targets related to sustainability issues, including absolute greenhouse gas emission reduction targets for 2030 and 2050, progress of activities, and achievements based on scientific evidence.

3. Description of the role and responsibilities of administrative, management, and control bodies with regard to sustainability issues.

4. Description of the firm's sustainability policies.

5. Information on adopted incentives.

6. Description of the due diligence process, highlighting actual or potential negative impacts associated with the firm's own activities and its value chain, actions undertaken to prevent, mitigate, remedy, or put an end to negative impacts and the outcomes of such actions.

7. Description of main risks and management methods.

8. Definition of relevant indicators for the provided information.

Furthermore, firms are required to disclose information on governance factors related to the following aspects: role of the firm's administrative, management, and control bodies; characteristics of internal control and risk management systems in relation to sustainability reporting and the decision-making process; corporate ethics and culture; exercise of political influence and lobbying activities; management practices, payment procedures, and quality of relationships with customers, suppliers, and communities.

2. Sustainability reporting standard: Setting and issue

Sustainability reporting standards intend to ensure the quality, understandability, relevance, comparability, and faithful representation of the reported information. The object of sustainability reporting standards provides for the disclosure of information in the following areas:

— environmental factors including mitigation and adaptation to climate change, reduction of greenhouse gas emissions, protection of water and marine resources, use of resources and circular economy; reduction of pollution; protection of biodiversity and ecosystems;

— social and human rights factors, including gender equality in treatment and opportunities, equal pay, training and skills development, employment and inclusion of people with disabilities, measures against violence and harassment in the workplace;

— working conditions, including secure employment and adequate wages, respect for working hours, social dialogue, freedom of association, the existence of works councils, collective bargaining, information, consultation, and participation rights of workers, work-life balance, health, and safety;

— respect for human rights, freedoms, and democratic principles established in the International Bill of Human Rights and other fundamental United Nations human rights conventions.

The International Financial Reporting Standards (IFRS) Foundation has approved the formation of the International Sustainability Standards Board (ISSB) in 2021. This body is dedicated to developing disclosure standards with the aim of providing, in the public interest, a comprehensive, comparable, and high-quality sustainability information base that takes into account the needs of investors and financial markets. The ISSB is supported in its work by the G7, G20, International Organization of Securities

Commissions (IOSCO), Financial Stability Board, African finance ministers, and the finance ministers and central bank governors of more than 40 countries. The Board's primary objectives are to develop global sustainability disclosure standards; fully and transparently inform investors and global capital markets; and facilitate interoperability with disclosures specific to jurisdictions and/or stakeholder groups. The ISSB integrates and makes more uniform, efficient, and connected the reporting systems defined by other bodies including the Climate Disclosure Standards Board (CDSB), the Task Force for Climate-based Financial Disclosures (TCFD), the Value Integrated Reporting Framework Reporting Foundation and SASB Standards, and the metrics developed by the World Economic Forum.

The adoption of consolidated and independent sustainability standards and reporting guidelines allows firms to establish objectives, priorities, and performance indicators. It also evaluates the effectiveness of corporate sustainability strategies and policies with respect to environmental performance (Grazhevska et al., 2020). The main issued standards include the IFRS Sustainability Disclosure Standards, SASB Standards, and Integrated Reporting. The IFRS standards and SASB Framework are complementary tools. The former specifies disclosure measures and metrics related to different sectors, while the latter provides guidance on the structure and content of reporting. On the other hand, Integrated reporting, driven by integrated thinking, encourages connected, concise, holistic, and strategically targeted communications.

IFRS—Sustainability Disclosure Standards. The ISSB in 2023 commenced the pubblication of the first two IFRS standards on sustainability. Specifically, they are IFRS S1 General Requirements for Disclosure of Sustainability-related Financial Information and IFRS S2 Climate-related Disclosures. The IFRS S1 and IFR S2 adoption is supported by the ISSB through the development of application guidelines and the establishment of the Transition Implementation Group (TIG). IFRS S1, becoming effective in January 2024, requires firms to provide information on risks and opportunities related to sustainability management to support decision-making processes. It outlines how firms manage, monitor, and report financial information and sustainability-related risks and opportunities across various areas. These include control systems and governance procedures for supervising risks and opportunities; strategy underlying risks and opportunities management; processes for identifying, evaluating and prioritizing risks and opportunities; and the performance achieved consistently with set objectives in relation to the risks and opportunities. IFRS S2, in force from January

2024, provided that IFRS S1 on General Requirements for Financial Reporting Related to Sustainability is also applied. The IFRS S2 goal is to disclose information on climate-related risks and opportunities, their evaluation and their impact on the firm's financial flows, and access to finance and financial prospects. Specifically, the standard applies to general, physical, and climate transition risks and opportunities related to it. This reporting activity generates information on governance processes, controls, and procedures to manage and oversee climate-related risks and opportunities; strategies for risks and opportunities management; identification and selection of priorities and their integration with the overall risk management process; and performance and expected results related to the climate with respect to targets to be achieved.

SASB—Sustainability Accounting Standards. The SASB standards (or industry standards) provide information on sustainability-related risks and opportunities that will impact cash flows, access to finance, or the cost of capital across 77 sectors. In 2022, the ISSB of the IFRS Foundation committed to developing and improving them. The standards setting process follows a rigorous and transparent path based on empirical evidence, the participation of firms, investors, and subject matter experts, and supervision by the SASB Standards Board. SASB standards play a crucial role in implementing the first two IFRS standards, S1 and S2. They contain disclosure topics, accounting parameters and associated technical protocols, and firm parameters relevant to each industry. An Application Guide, as an integral part of the standards, has been developed to improve their implementation. Standards and Application Guide are ruled by other important documents such as the SASB Conceptual Framework and the SASB Procedural Rules.

IR—Integrated Reporting. The IR is a document provided by the International Integrated Reporting Council. The IIRC is a nonprofit organization promoted by A4S, an association established by the Prince of Wales and the GRI—Global Reporting Initiative. In 2021, the IIRC and the Sustainability Accounting Standards Board (SASB) merged to establish the Value Reporting Foundation to jointly develop the Integrated Reporting Framework. The Value Reporting Foundation is a nonprofit organization operating internationally with the aim of assisting firms in developing a shared understanding of enterprise value and how it is created, preserved, or eroded over time. IR represents a value generation model that provides tools and methodologies to innovate the firm's business model. It has various purposes, including improving the quality of information on financial capital; promoting efficient approaches to corporate reporting; increasing responsibility and financial management, productive, intellectual, human, social, relational, and natural capitals; and supporting the decision-making process in integrating and finalizing actions toward value creation. IR is useful for meeting stakeholders' information needs on the organization's ability to create value over time. Moreover, the report is designed to strike a balance between flexibility and requirements in dynamic organizations and to foster a level of comparability of relevant information among different organizations. To favor the IR adoption, the International Integrated Reporting Framework has been developed. It takes a principles-based approach rather than specifying performance indicators, measurement methods, and types of information. It identifies seven Guiding Principles on how information should be prepared and presented, including Strategic Focus and Future Orientation, Connectivity of Information, Stakeholder Relations, Materiality, Conciseness, Reliability and Completeness, Consistency, and Comparability.

The principled approach allows firms to improve the quality of information available to providers of financial capital and integrate different lines of corporate reporting more cohesively. Published in 2013, the Integrated Reporting Framework has undergone revisions following extensive consultation with individuals and jurisdictions. The revisions to the Integrated Reporting Framework, published in January 2021, demonstrated the robustness and appropriateness of integrated thinking and the principles that underpin it.

3. The sustainability reporting standards list

Firms are implementing measures to ensure their operations align with international sustainability standards. At the international level, certain reporting standards are widely recognized, signaling the direction in which the regulatory process of social and environmental reporting is evolving and the trend toward increasingly shared and recognized reporting standards.

International standards provide firms with guidelines and models to increase productivity and efficiency while simultaneously promoting positive environmental and social impacts. Adherence to these measures serves to monitor the firm's performance in terms of reducing the carbon footprint, utilizing renewable energy, and employing eco-sustainable resources.

Sustainability standards, whether voluntary or private, present similarities in main objectives and certification procedures, but they differ in terms of historical development, target audience, geographical spread, and focus on environmental and social issues. Institutions and standardization bodies promote standardization in reporting to achieve universal standards for sustainability reporting.

Table 4.1 shows the principal 2023 international sustainability standards used by firms to manage their operations responsibly. It incudes EU CSRD (ESRS), UNFSS, IIRC, TCFD, ISSB (IFRS), CDP, B Corp, GRI and SASB (Value Reporting Foundation), and ISO 14001.

BCORP— B Lab's standards address social impact, sustainability, and ESG topics. They define the best social, environmental, and governance practices for businesses, particularly for small and medium enterprises. Standards are governed by B Lab's Standards Advisory Council, an independent, global, multistakeholder group specializing in responsible and sustainable business. They include input from external stakeholders, working groups, and advisory groups. To obtain the B Corp private certification, firms have to achieve a score of 80 in the B impact assessment, which include:

— B Impact Assessment, the evaluation of the positive impact of a firm's performance over a 12–18-month period on workers, communities, customers, suppliers, and the environment.
— Risk Standards, an assessment of eligibility for B Corp Certification based on an examination of potentially negative impacts associated with the industry and other practices.
— Multinational Company Standards and Baseline Requirements, and additional baseline requirements for large corporations defined as parent companies generating more than $5 billion in annual revenue.
— Engaging stakeholders to define new and emerging topics and best practices.
— Identification of priority areas for improvement.
— Continuous research and development.
— Testing and data analysis.
— A 60-day public comment period.

Benefit governance considers the interests of various stakeholders including investors, customers, workers, suppliers, communities, and environment.

CDP—The CDP (previously Carbon Disclosure Project) is a nonprofit charity that oversees the global disclosure system on the environmental impacts of firms, investors, organizations, cities, states, and regions. CDP is particularly suitable for medium and large companies interested in climate, supply chain, forest, and water topics. CDP is considered as the gold standard of environmental reporting due to its comprehensive and rich dataset on the actions of firms and cities. Established in 2000 as the "Carbon Disclosure

Table 4.1 2023 International sustainability standards overview.

Sustainability standard	Topic	Best suited for
B Corp https://www.bcorporation.net/en-us/	Best social, environmental, and governance practices	Small and medium enterprises
CDP—Carbon Disclosure Project https://www.cdp.net/en	Climate, supply chain, forest and water	Medium and large companies
ESRS—EU Sustainability Reporting Standards https://www.efrag.org/lab6	Sustainability and environmental reporting, including governance issues, climate change, biodiversity, and human rights	Companies with 500+ employees and/or that are publicly traded
IIRC—International Integrated Reporting Council https://www.integratedreporting.org/	Finance, ESG, sustainability	Investors
GRI—Global Reporting Initiative https://www.globalreporting.org/	Sustainability	Any type of organization
ISSB (IFRS)—International Sustainability Standard Board https://www.ifrs.org/groups/international-sustainability-standards-board/	General sustainability accounting, risks, and opportunities	Medium and large companies
SASB—Sustainability Accounting Standards Board https://sasb.org/	ESG financial risk	Large companies
TCFD—Task Force on Climate-related Financial Disclosures https://www.fsb-tcfd.org/	ESG and climate financial risk	Large companies
UNFSS—United Nations Forum on Sustainability Standards https://unfss.org/home/about-unfss/	ESG and sustainability reporting	All industries
ISO 14001—International Organization for Standardization https://www.iso.org/iso-14001-environmental-management.html	ESG and environmental management systems	Small and large companies

Source: From Author's elaboration.

Project" to make climate impact data public, CDP has expanded the scope of environmental disclosure to include deforestation and water security for cities, states, and regions. The 2021 strategy involves further extension into new countries and areas, such as biodiversity, plastics, and oceans. Today, *CDP Global* is an international nonprofit organization composed of CDP Worldwide Group, CDP North America, Inc., and CDP Europe AISBL. It has regional offices and local partners in 50 countries. Companies, cities, states, and regions in over 90 countries disclose through CDP annually, with scores of reporting firms published on CDP website. The primary audience of stakeholders is prevalently composed of investors and suppliers.

ESRS—The European Sustainability Reporting Standards, adopted by the European Commission, represents an important tool for the transition toward a sustainable, transparent, and responsible economy. The main topics are sustainability and environmental reporting, required for EU companies with 500+ employees and/or publicly traded. The first set of ESRS was adopted in July 2023 and is mandatory for companies subject to the Corporate Sustainability Reporting Directive (CSRD). Compliance with them allows firms to access sustainable finance, contributing to the attainment of the 2030 agenda objectives. The ESRS scope is broad, covering environmental, social, and governance issues, including climate change, biodiversity, and human rights. These standards offer investors valuable information into social, environmental, and governance matters, facilitating a better understanding of the impact that the companies they invest in have on sustainability. They also consider interventions from the International Sustainability Standards Board (ISSB) and the Global Reporting Initiative (GRI) to ensure interoperability between European and global standards, minimizing the redundancy of disclosed information. The ultimate goal is to align sustainability reporting more closely with financial accounting and reporting standards. The primary stakeholders are prevalently represented by regulators and other EU institutions.

IIRC—The International Integrated Reporting Council was created as a consultative body with the aim of integrating the reporting required by the IASB (International Accounting Standards Board) and the ISSB (International Sustainability Standards Board) and provide guidance on how to consider the principles of the Integrated Reporting Framework in corporate projects.

Established in August 2010, the IIRC was responsible for defining a global reference framework regarding the communication procedures to be followed for value creation over time. The body is made up of a global coalition of regulators, investors, businesses, standard setters, accounting professionals, academia, and NGOs. It is currently incorporated into the Integrated Reporting and Connectivity Council of the IFRS Foundation, a nonprofit public interest organization committed to developing high-quality sustainable accounting and disclosure standards globally. Under this new composition, the IIRC continues to consult with the IFRS Foundation Trustees, the International Accounting Standards Board (IASB), and the International Sustainability Standards Board (ISSB). The topics covered include finance, ESG, and sustainability, primarily addressed to investors. The purpose of this framework is to establish Integrated Sustainability guiding principles and content elements of the integrated report. It provides coherent and connected financial reporting guidance and packages on different areas, such as the integration of reporting required by the IASB and the ISSB and how the principles and concepts of the Integrated Reporting Framework are applied by the IASB and the ISSB. The main stakeholder audience is represented by regulatory bodies and accountancy firms.

ISSB (IFRS)—IFRS S1 and IFRS S2 were issued by the International Sustainability Standards Board (ISSB) to introduce sustainability-related disclosures to capital markets around the world. These standards require the use of a common language to disclose the impact of climate-related risks and opportunities on a firm's prospects. The main topics are, therefore, General Sustainability Accounting, Risks, and Opportunities and are particularly well suited for medium and large companies. The primary stakeholder audience are investors, CFOs, and finance professionals. Through IFRS S1 and IFRS S2essi, the objective is to enhance trust in company information on sustainability and support investors in making informed investment decisions.

SASB—The Sustainability Accounting Standards Board supports firms in disclosing relevant information on ESG financial risk. SASB develops nonfinancial sustainability reporting standards to monitor and communicate the most financially relevant ESG performance areas and metrics for investors. SASB standards vary by industry and are available for 77 industries. They identify risks and opportunities related to sustainable investments to assess their impact on an entity's cash flows, access to finance, and cost of capital in the short, medium, or long term. In 2021, the Value Reporting Foundation was created through the merger of SASB and IIRC (International Integrated Reporting Council). Since 2022, SASB standards are regulated by the International Sustainability Standards Board (ISSB) of the IFRS Foundation, to provide an integrated reporting framework linking sustainability reporting with financial reporting. These standards are suitable for large companies, with investors being the primary stakeholder audience.

TCFD—Task Force on Climate-related Financial Disclosures was created to develop specific information on financial activities related to climate risk. The standards, focused on ESG and climate financial risk issues, are used by large companies. The primary audience includes investors, shareholders, financiers, and insurance underwriters to support them promptly and accurately in capital allocation decisions. TCFD was created by the Financial Stability Board (FSB) to formulate a pillar-based recommendations framework, enhancing market transparency. The use of a common language in both financial and nonfinancial markets, as part of strategic planning and risk management processes, allows for a more reliable assessment of the financial implications of the transition to a sustainable economy. The disclosure recommendations are structured into four interconnected thematic areas, such as governance, strategy, risk management, metrics, and objectives.

UNFSS—The United Nations Forum on Sustainability Standards is an initiative with a committee made up of six United Nations Agencies (Food and Agriculture Organization—FAO, the International Trade Centre—ITC, UN Environment, UN Industrial Development Organization—UNIDO, United Nations Economic Commission for Europe—UNECE, and the UN Conference on Trade and Development—UNCTAD). The topics covered include ESG and sustainability Reporting for all industries, with the primary stakeholder audience being National and International bodies.

The Forum, being neutral and independent, takes into consideration specific sectors such as forestry, agriculture, mining, fishing, biodiversity, and the reduction of greenhouse gas emissions. It primarily targets developing countries, assisting them in accessing global markets, reducing local poverty, and protecting ecosystems and natural resources. The voluntary sustainability standards (VSS) developed by it are applied by producers, suppliers, and traders to certify that their economic activities do not harm people, workers, and environment. VSS aims to protect communities and territories by promoting proactive and strategic dialogue on policies and meta-governance issues. UNFSS publishes a semiannual Flagship Report on topics relevant and useful for understanding issues related to VSS opportunities and challenges for developing economies.

ISO 14001—The International Organization for Standardization is an independent international organization made up of 169 national bodies.

Through this broad participation, the organization shares knowledge and develops voluntary, consensus-based international standards on innovation to address global sustainability challenges. ISO refers to ESG and environmental management systems topics and its standards are suitable for small and large companies. ISO 14000 comprises a group of standards related to environmental management with the aim of assisting organizations in reducing the negative impact of their activities on the environment by complying with laws and regulations and improving their environmental performance. The ISO 14001 standards, developed by the ISO Technical Committee ISO/TC 207 and its various subcommittees, establish criteria for creating an effective environmental management system through certification. They are designed for any type of organization and focus on specific areas such as audits, communications, labeling and lifecycle analysis, environmental challenges, and climate change. ISO 14001 represents the international standard for environmental management systems and is supported by other standards (for example the ISO 14002, ISO 14004, ISO 14005, ISO 14006, ISO 14007, ISO 14008, ISO 14009, ISO 14053 series and in progress development standards ISO 14002-3 and ISO 14002-4, ISO 14054). The primary stakeholder audience includes investors and regulators.

4. The Global Reporting Initiative

The Global Reporting Initiative (GRI) framework has been developed within the several models and guidelines shared at an international level. It provides tools, procedures, and criteria for producing the sustainability report, enabling firms to account for their results obtained in a responsible and transparent way. The logic followed by GRI is to create a global report that helps firms measure environmental performance without limiting their profitability while enhancing their reputation and industrial relations.

Global reporting identifies the phases to follow in preparing the document, including the preliminary identification of the firm's stakeholders and their information needs; collection of information independent of partisan interests and verifiable; selection of qualitative-quantitative data expressing the ethical, social, and environmental value of activities carried out; and transparency in communicating the logical process followed in drawing up the sustainability report. According to GRI framework, the report identifies the following areas:

- Definition of firm through clarification of mission and organizational models adopted;

- Environmental and social reporting with prior identification of stakeholders;
- Identification of correspondences with programs to evaluate the activity's compliance with the pursuit of social responsibility.

The global report integrates with financial statements, establishing a connection between economic and social and environmental data as well as between decision-making processes that contain economic aspects and their effects in the ethical, social and environmental dimensions.

To set and measure the areas indicated above, firms adopt GRI standards to evaluate performance on the basis of specific indicators, aligning themselves with international ESG goals. These standards represent global best practices for publicly reporting a range of economic, environmental, and social impacts. Standards-based sustainability reporting provides information on an organization's positive or negative contributions to sustainable development.

GRI standards allow firms to have a common global vision and bring together organizations worldwide through the adoption of a shared language and concepts applicable in every sector and country. Firms evaluate their position in the sustainable field through a self-declaration based on reference standards, requesting specific certification from GRI. Standards support sustainability reporting, identify risks related to business sectors, and drive firms towards changes in selected areas of analysis by monitoring risk factors. They constitute shared, multidimensional, and unique parameters applicable to all organizations across different sectors.

The full set of GRI Standards is listed below.

- GRI 1: Foundation 2021
- GRI 2: General Disclosures 2021
- GRI 3: Material Topics 2021
- GRI 11: Oil and Gas Sector 2021
- GRI 12: Coal Sector 2022
- GRI 13: Agriculture, Aquaculture, and Fishing Sectors 2022
- GRI 201: Economic Performance 2016
- GRI 202: Market Presence 2016
- GRI 203: Indirect Economic Impacts 2016
- GRI 204: Procurement Practices 2016
- GRI 205: Anticorruption 2016
- GRI 206: Anticompetitive Behavior 2016
- GRI 207: Tax 2019
- GRI 301: Materials 2016

- GRI 302: Energy 2016
- GRI 303: Water and Effluents 2018
- GRI 304: Biodiversity 2016
- GRI 305: Emissions 2016
- GRI 306: Effluents and Waste 2016
- GRI 306: Waste 2020
- GRI 308: Supplier Environmental Assessment 2016
- GRI 401: Employment 2016
- GRI 402: Labor/Management Relations 2016
- GRI 403: Occupational Health and Safety 2018
- GRI 404: Training and Education 2016
- GRI 405: Diversity and Equal Opportunity 2016
- GRI 406: Nondiscrimination 2016
- GRI 407: Freedom of Association and Collective Bargaining 2016
- GRI 408: Child Labor 2016
- GRI 409: Forced or Compulsory Labor 2016
- GRI 410: Security Practices 2016
- GRI 411: Rights of Indigenous Peoples 2016
- GRI 413: Local Communities 2016
- GRI 414: Supplier Social Assessment 2016
- GRI 415: Public Policy 2016
- GRI 416: Customer Health and Safety 2016
- GRI 417: Marketing and Labeling 2016
- GRI 418: Customer Privacy 2016
- GRI Standards Glossary.

The GRI Standards full set is a modular system of interconnected standards. The complete set can be grouped into three sets of Standards, namely the GRI Universal Standards (GRI 1, GRI 2, and GRI 3), applicable to all organizations, the GRI Sector Standards applicable to specific sectors, and the GRI Topic Standards, providing relevant information for specific material topics (Global Reporting Initiative, 2021).

The Universal Standards GRI 1, GRI 2, and GRI 3 serve as guiding principles for sustainability reporting. Specifically, GRI 1 introduces the purpose and system of the GRI Standards, explains key concepts for sustainability reporting, and specifies reporting principles in accordance with the GRI Standards. GRI 2 contains indications on the information to be provided to implement good reporting practices and on the organizational and governance details relating to the application context. GRI 3 represents a guide for determining, listing, and managing material topics.

The Sector standards offer information on probable material themes and their definition. The Topic standards contain useful information for detecting impacts in relation to particular topics. The adoption of a reporting system in compliance with the GRI Standards allows organizations to provide a complete picture of their most significant impacts on the economy, environment, and community to realize informed and conscious assessments and decisions. Reporting in accordance with the GRI Standards involves the following steps.

— Requirement 1: Apply the reporting principles
— Requirement 2: Report the disclosures in GRI 2: General Disclosures 2021
— Requirement 3: Determine material topics
— Requirement 4: Report the disclosures in GRI 3: Material Topics 2021
— Requirement 5: Report disclosures from the GRI Topic Standards for each material topic
— Requirement 6: Provide reasons for the omission of disclosures and requirements that organization cannot comply with
— Requirement 7: Publish a GRI content index
— Requirement 8: Provide a statement of use
— Requirement 9: Notify GRI.

Compliance with the nine requirements is necessary to report according to the GRI Standards. The organization that complies with these requirements can declare that it has prepared the information reported with the formula at the end of the section "Reporting with reference to the GRI Standards."

5. ESG pillars and corporate sustainability reporting

Firms and organizations can choose among different standards and reporting models to define their Environmental, Social, and Governance (ESG) parameters and accounting system to detect and communicate sustainability results (Meseguer-Sánchez et al., 2021). The term ESG refers to the three fundamental pillars used to measure the level of firms' sustainability. ESG was first introduced in 2004 in the "Who Care Wins" report that highlights the importance of the interconnection of social, environmental, and governance issues to improve the quality of corporate management. The ESG pillars reflect the firm's commitment to antiwaste policies, reduction of polluting emissions, fight against deforestation, biodiversity care, limitation of the chemical substances used in production activities, and responsible use of natural resources. In line with this commitment, governance takes on a central role in defining rules, principles, code of ethics, risk management, and policies consistent with the ESG approach. The Action Plan on Financing Sustainable, published in 2018 by the European Commission, aims to encourage investments in sustainable projects that meet ESG criteria. It introduces an EU taxonomy to classify sustainable economic activities that fall within financing plans, such as mitigation and adaptation to climate change; protection of waters and marine resources; transition to a circular economy; waste management. Additionally, it defines information obligations and requirements on ESG factors to aid investor decision-making.

The ESG approach is strictly related to the Corporate Social Responsibility (CSR) approach, which expresses how firms act and interact with their environment to have a positive impact on society (Diez-Cañamero et al., 2020; Freeman et al., 2006). The aim of maximizing profit is linked to compliance with ESG objectives and a corporate culture where social and environmental responsibility are integral components of governance. The impact of CSR on corporate sustainability has been analyzed in the literature through social and economic performance, as corporate sustainability considers ethical issues and the possibility of creating jobs and promoting community development (Carroll, 2015; Phillips et al., 2020; Stutz, 2018).

Corporate Sustainability Reporting contains results on the performance of companies, governments, investors, and NGOs with respect to sustainability. It has a dual function. On the one hand, it represents the basic tool for firms to evaluate the impact on various areas of sustainability, including environmental footprint, polluting emissions, and sustainable production processes. On the other hand, it serves to publicly share risks, results, performances, and environmental practices implemented within firms, allowing stakeholders to stay informed about the company's activities and make more informed decisions. The adoption of the sustainability report by firms is a tool for improving brand reputation and attracting new capital and financing. It is a voluntary disclosure tool used as a strategic communication lever with stakeholders, in addition to the mandatory disclosure of which the financial statements are an expression. Sustainability reporting fulfills the duty of accountability, in the dual sense of providing useful data to support internal company decisions and explaining the relationships between mission, governance, and accountability on an external level.

In summary, sustainability reporting is a system for measuring, disclosing, and being accountable for social, environmental, and governance performance using specific metrics and incorporating sustainability into corporate strategy and policies. It expresses the organization's principles, values, and commitment in terms of future commitments and objectives.

The sustainability report contains social and environmental accounting in a single document to provide a comprehensive overview of the economic activity's impact on the society. The nonfinancial statement includes a section called "*calculations*" dedicated to monetary and nonmonetary measures and summary indices, and a textual section, called "*narratives*" for the interpretation and understanding of data and results (Burritt & Schaltegger, 2010). To make data and information more useable and accessible, images and graphic content can be used in line with impression management. The report can be drawn up in two ways. The first "*from talk to walk*" or "*outside-in*" provides the path from communication to management to have immediate feedback from the interlocutors. It is a market-oriented approach that presents the risk of anticipating actions and objectives that may not be achieved, potentially damaging the firm's image and credibility. The second mode "*from walk to talk*" or "*inside-out*" involves the path from management to communication. It starts from the internal policies adopted and then communicates them externally in a transparent way, allowing stakeholders to grasp the solidity of the processes.

Over the last decade, sustainability reporting and disclosure have grown rapidly (https://www.brightest.io/sustainability-reporting-standards). According to GRI (Global Reporting Initiative) 2019 data, almost 90% of companies with the highest level of market capitalization have adopted and published the sustainability report.

Unlike traditional financial accounting documents, sustainability reporting is not yet well consolidated in terms of transparency, consistency, and linearity. This is confirmed by the presence of over 600 different sustainability reporting standards, sector-specific initiatives, or voluntarily adopted company policies. Sustainability reporting is complex, articulated, and not easily amenable to a uniform and organic framework. Firms that intend to adopt sustainable reporting face numerous regulations, directives, standards, and procedures, risking duplication of activities and information. Consequently, sustainability reporting is a partly discretionary choice and is contextualized to the sector and the methods of measuring and communicating performance.

In selecting sustainability standards, firms take into account their diffusion in the sector, their consistency with the activity carried out, and the extent of the investments, priorities, and actions that they are called upon to carry out. Some criteria guide the validity, significance, and reliability of the sustainability report. Among the most important are:

— stakeholder inclusiveness, through the stakeholders identification and methods of responding to their expectations and requests;
— sustainability context delimiting the firm's ESG actions and their implications on the reference context;
— identification of material topics on which the sustainability strategy is based to understand the organization's ability to create value in the direction undertaken and to evaluate the credibility of the methodology used in terms of identification, selection, and significance of social, environmental, and economic items;
— verification of results to evaluate the internal and external solidity and credibility of the objectives, expected results, and possible implications;
— completeness and clarity of information and data to guarantee the quality of the report and represent the positive and negative aspects of performance in a balanced and reliable way. This allows both governance and stakeholders to formulate a critical assessment of sustainable management;
— impartiality of information offered, avoiding omissions, selections, and representations that could lead to errors and misleading interpretations.

These criteria guide firms in defining and representing the sustainable business model in an organic, solid, and coherent way with the value creation process, the representation of the activities impacts, the analysis of materiality, the description of the role and weight of stakeholders, and the analysis of corporate ESG performance.

References

B Corp. https://www.bcorporation.net/en-us/.

Burritt, R. L., & Schaltegger, S. (2010). Sustainability accounting and reporting: Fad or trend? *Accounting, Auditing and Accountability Journal, 23*(7), 829–846.

Carroll, A. B. (2015). Corporate social responsibility: The centerpiece of competing and complementary frameworks. *Organizational Dynamics, 44*(2), 87–96. April–June 2015.

CDP—Carbon Disclosure Project. https://www.cdp.net/en.

Das, J. K., Taneja, S., & Arora, H. (Eds.). (2021). *Corporate social responsibility and sustainable development: Strategies, practices and business models* (1st ed.). New York, NY, USA: Taylor and Francis Group. ISBN 978-0-367-27304-0.

Diez-Cañamero, B., Bishara, T., Otegi-Olaso, J. R., Minguez, R., & Fernández, J. M. (2020). Measurement of corporate social responsibility: A review of corporate sustainability indexes, rankings and ratings. *Sustainability, 12*, 2153.

Directive 2014/95/EU of the European Parliament and of the Council of 22 October 2014 on non-financial information disclosure and diversity. Retrieved from https://eur-lex.europa.eu/legal-content/EN/TXT/?uri=celex%3A32014L0095.

Directive (EU) 2022/2464 of the European Parliament and of the Council of 14 December 2022 corporate sustainability reporting. Retrieved from https://eur-lex.europa.eu/legal-content/EN/TXT/?uri=CELEX%3A32022L2464.

ESRS—EU Sustainability Reporting Standards. https://www.efrag.org/lab6.

Fallah Shayan, N., Mohabbati-Kalejahi, N., Alavi, S., & Zahed, M. A. (2022). Sustainable development goals (SDGs) as a framework for corporate social responsibility (CSR). *Sustainability, 14*, 1222.

Freeman, R. E., Velamuri, S. R., & Moriarty, B. (2006). Company stakeholder responsibility: A new approach to CSR. *Business Roundtable Institute for Corporate Ethics, Bridge Papers, 19*.

Global Reporting Initiative. (2021). GRI Standards: Universal, Sector and Topic Standards. https://www.globalreporting.org/media/wtaf14tw/a-short-introduction-to-the-gri-standards.pdf. (Accessed 30 October 2023).

Grazhevska, N., & Mostepaniuk, A. (2020). Ecological components of corporate social responsibility: Theoretical background and practical implementation. *Journal of Environmental Management & Tourism, 11*(5), 1060−1066.

GRI—Global Reporting Initiative. https://www.globalreporting.org/.

Hussain, R. I., Bashir, S., & Hussain, S. (2020). Financial sustainability and corporate social responsibility under mediating effect of operational self-sustainability. *Frontiers in Psychology, 11*, 550029.

IIRC—International Integrated Reporting Council. https://www.integratedreporting.org/.

ISO 14001—International Organization for Standardization. https://www.iso.org/iso-14001-environmental-management.html.

ISSB (IFRS)—International Sustainability Standard Board. https://www.ifrs.org/groups/international-sustainability-standards-board/.

Latapí Agudelo, M. A., Jóhannsdóttir, L., & Davíðsdóttir, B. (2019). A literature review of the history and evolution of corporate social responsibility. *International Journal of Corporate Social Responsibility, 4*(1), 1−23.

Meseguer-Sánchez, V., Gálvez-Sánchez, F. J., López-Martínez, G., & Molina-Moreno, V. (2021). Corporate social responsibility and sustainability. A bibliometric analysis of their interrelations. *Sustainability, 13*, 1636.

Phillips, R., Schrempf-Stirling, J., & Stutz, C. (2020). The past, history, and corporate social responsibility. *Journal of Business Ethics, 166*(2), 203−213.

SASB—Sustainability Accounting Standards Board. https://sasb.org/.

Stutz, C. (2018). History in corporate social responsibility: Reviewing and setting an agenda. *Business History, 63*, 175−204.

TCFD—Task Force on Climate-related Financial Disclosures. https://www.fsb-tcfd.org/.

UNFSS—United Nations Forum on Sustainability Standards. https://unfss.org/home/about-unfss/.

Zhao, F., Kusi, M., Chen, Y., Hu, W., Ahmed, F., & Sukamani, D. (2021). Influencing mechanism of green human resource management and corporate social responsibility on organizational sustainable performance. *Sustainability, 13*, 8875.

CHAPTER FIVE

Strategies for sustainable innovations in fashion firms

1. SDG 9 for sustainable fashion

The sustainable development goals (SDGs) outlined in the 2030 Agenda represent the largest action program for sustainable development, encompassing the fashion industry as well. The SDGs that guide the industrial and commercial policies of the fashion sector concern several interconnected areas, including health and well-being, quality education, renewable energy, decent work and economic growth, sustainable cities and communities, responsible consumption and production, climate, water, chemicals, and raw materials (Papamichael et al., 2022; Thakker & Sun, 2023). The most active countries in the search for sustainable products are Denmark, which emerges as a leader in eco-sustainable productions, while Australia stands out for its demand for sustainable sportswear, swimwear, and denim and for ethical fashion. Germany, Spain, and France exibit a preference for vegan, cruelty-free, and recycled materials. Italy recorded a 20% increase in searches compared to the previous year, with a focus on ecological fur and recycled fashion (De la Motte & Ostlund, 2022; Sinha et al., 2023). In particular, Goal 9 focuses on a sustainable and inclusive industry with a significant impact on employment, R&D, and gross domestic product (GDP) in both developed and developing countries. To pursue Goal 9, firms are called upon to renew infrastructure and make production processes more efficient through technological empowerment and eco-friendly industrial processes (Jovanovich, 2022). Greater financial, technical, and technological support is envisaged for firms in developing countries to support the renewal of productive structures and encourage the integration of small and medium-sized enterprises into the markets.

To assess the impact of technological investments on achieving Goal 9 (UNECE, 2023), several useful indicators include total expenditure and investments allocated for environmental protection; extent and nature of investments in sustainable infrastructure and services; type, size, and capacity of innovative sustainable technologies; and quantification of economic

Being a Sustainable Firm
ISBN: 978-0-443-14062-4
https://doi.org/10.1016/B978-0-443-14062-4.00002-7

and social value generated by the modernization process. Technological renewal is crucial for the fashion industry, which contributes significantly to the environmental pollution rate.

According to recent data (UNECE, 2023), the fashion industry is responsible for producing 10% of global carbon emissions, 20% of global wastewater, 24% of insecticides, and 11% of pesticides used for cotton cultivation. In addition, the textile industry is considered a major source of plastic pollution, with approximately half a million tons of plastic microfibers ending up in the oceans annually during the washing of fabrics, such as polyester, nylon, or acrylic. In the fashion industry, precarious and unsafe working conditions are sometimes present due to intense work schedules, low wages, the use of toxic substances in production processes, and situations of child labor often found in developing countries. These aspects were recently highlighted in the UNECE Regional Forum promoted in collaboration with the United Nations Alliance for Sustainable Fashion. The Forum brought together Connect4Climate, the World Bank Group, the International Labor Organization, the ITC Ethical Fashion Initiative, the United Nations Development Programme, the United Nations Economic Commission for Europe, the UN Environment, the UN Global Compact, and the United Nations Office Unite for partnerships and key representatives of the fashion industry to share experiences and solutions for achieving sustainability goals. In recent years, the fashion industry has been engaging in the implementation of sustainable business models and in the redesign of value chains according to the circular economy model. In this context, key elements are the traceability and transparency of supplies, ethical fashion, slow fashion, garment circularity, and second-hand recovery.

Despite having programs focused on improving worker conditions, respect for human and environmental rights, firms may sometimes provide insufficiently transparent information or incomplete data on their efforts. Greenwashing, for example, is a communication strategy implemented by firms to promote programs for environmental sustainability and strengthen their positioning in terms of positive environmental impact (Mukendi et al., 2020; Saxena & Khandewal, 2010; Yazdanifard & Mercy, 2011). However, these efforts are often not adequately supported by data and information on improvements in production processes. Especially, fast fashion brands tend to adhere to awareness campaigns or sustainability programs leveraging emotional aspects of consumers without providing feedback on the practices carried out. In this practice, the issue of sustainability has a marginal value or is limited to a mere communication environmental activity, not properly aligned with the sustainability goals. The sustainable capsules are another example of activities that target ethical consumers, through specific collections focusing on brand values, regardless of any actual content.

2. Circularity in the fashion life cycle

In recent years, the issue of sustainability has become one of the main priorities for fashion firms, as they are committed to pursuing emission reduction objectives, waste disposal of clothes, and facing climate change. Fashion firms are called to act responsibly to reduce the excessive waste generated by clothing quickly going out of fashion. This chapter examines strategies and approaches that fashion firms can draw inspiration from for moving from traditional production models to circular systems.

Firms are engaged to revise the entire life cycle of their products since environmental problems are largely attributable to global disposal practices. Overall, 85% of fabrics end up in landfills or are incinerated rather than recycled. Every year, in the United Kingdom, around one million tons of new clothes are bought, with half of them ending up in landfills and the other half being recycled or sold in vintage or charity shops. In the United States, about 23.5 million tons of clothes are purchased every year, with 10% ending up in the second-hand market while the remainder in landfills or given to charity (Thorisdottir & Johannsdottir, 2019). The sector is increasingly focused on environmental problems caused by the short life cycle of clothing. This brevity is linked both to corporate strategies that prompt sudden changes in fashion and to market variations, sensitive to new proposals. Fashion firms are moving towards sustainable strategies based on the principles of the circular economy. The 2020 Sustainability Report by Global Fashion Group (2020) highlights an increase in searches for sustainability-related keywords. Ethical fashion, slow fashion, traceability and transparency of the supply chain, and circularity are the keywords on which the current orientation toward sustainability is based. The report also declines the issue of corporate responsibility toward people and the planet in terms of Ethical Trade, Environmental Operations, Community, People, and Governance, Risk, and Compliance.

The sustainability strategy involves both corporate and business levels to get a competitive advantage grounded on productivity and reputation. This strategy is inspired by the principles of ethical values and respect for the planet (Avelino & Wittmayer, 2016; Loorbach & Wijsman, 2013; Markard et al., 2012), revisiting raw materials, human resources, and environmental resources in a circular key. Some fashion firms, such as H&M, Louis Vitton, and Hugo Boss, have successfully implemented sustainable business models, rethinking the different stages of fashion product life in terms of circularity.

The main challenge for fashion firms is to integrate sustainability with traditional business models (Dwivedi et al., 2019; Zimmer et al., 2016), applying the circular model to the product life cycle, to improve the efficiency of production processes, optimize the use of raw materials, manage production waste and reduce warehouse stocks, extend the product life duration through the reuse, repair, and recycling of the clothing, and guarantee the traceability and transparency of the supply chain.

The implementation of the circular model requires investments in plants with high fixed costs, leading most of the time to negative financial and economic performance in the short term. It also requires concrete actions, experimentation, and the use of digital technologies facilitating the application of the circular model to the green fashion life cycle. Among these, technologies based on modularity, versatility, and adaptability play a crucial role in extending the life cycle of clothing by enabling the reuse and recycling of products. Digitization in the design and eco-design phase makes use of digital and 3D printing technologies to limit fabric and materials waste. Blockchain serves as a solution for tracing the entire textile supply chain, ensuring information transparency and control throughout the supply chain (Dey & Cheffi, 2013; Klassen & Johnson, 2004; Martel & Klibi, 2016). The spread of digital platforms and apps provides an ideal marketplace for the second-hand sale of green fashions, attracting consumers who are attentive to sustainability issues.

3. Directives, regulations, and certifications for sustainable textiles

A significant number of textile firms have signed the Sustainable Fashion Charter with the aim of reducing greenhouse gas emissions produced by the entire fashion chain by 2030. Interventions in several areas have been adopted, including the reduction of coal-based energy production sources, the use of low carbon emissions transportation, the utilization of eco-sustainable materials, and the development of partnerships for designing and implementating circular economy systems.

In the field of textiles, numerous directives, regulations, and certifications are in place to guide fashion firms in managing and monitoring sustainability. Regulation no. 1007/2011 establishes rules on the use of textile fibers, labeling, and marking of textile products to improve traceability of clothes and accuracy of information (European Parliament and of the Council, 2011). Additional EU regulations of interest to the textile and clothing sector are:

- Directive 2001/95/EC concerning general product safety;
- Regulation (EC) no. 850/2004 on persistent organic pollutants -POPs (European Parliament and of the Council, 2004);
- Regulation (EC) no. 1907/2006 REACH (Registration, Evaluation, Authorization of Chemicals), concerning the registration, evaluation, authorization, and restriction of chemicals and the establishment of the European Chemicals Agency (European Parliament and of the Council, 2006);
- Regulation (EC) no. 66/2010 related to the European Union ecological quality label (EU Ecolabel);
- Regulation (EU) no. 528/2012 concerning the availability on the market and use of biocides (European Parliament and of the Council, 2012).
  Recent directives and regulations are the following:
- The Extended Producer Responsibility (EPR). These rules (OECD, 2001) consider the producer's responsibility for textile products to reduce waste and limit environmental impacts of products.
- The Corporate Sustainability Due Diligence Directive (CSDDD). It ensures that environmental and human rights due diligence is carried out within companies' supply processes (European Commission, 2022).
- The Corporate Sustainability Reporting Directive (CSRD). The Directive (EU) 2022/2464 (European Parliament and of the Council, 2022) on sustainability reporting also extends to the context of the fashion industry with the goal of reporting, communicating and monitoring sustainability dimensions.
- Eco-Design for Sustainable Products Regulation (European Commission, 2009). This legislation focuses on eco-design to improve outcomes in the areas of resource use, repairs, and circularity. Based on the indications contained in the regulation, the product is provided with a "passport" containing useful information, enabling consumers to make informed and safe decisions.
- Textile Exchange Recycled Claim Standard (RCS) refers to recycled products and waste sorting, encompassing those produced both before and after consumption. It verifies the presence and quantity of recycled material in a final product through a chain of custody verification (CCPB Control and Certification, 2023).

- Organic Content Standard (OCS) checks the organic content of a product. For example, if the raw material has been certified as organic, its organic content is traced throughout the production and transformation activities process (Textile Exchange, 2023).
- Global Traceable Down Standard (Global TDS) and the Responsible Down Standard (RDS) guarantee that the down used in clothing products comes from responsible sources respecting animal welfare (NSF, 2023).
- Responsible Wool Standard (RWS) safeguards the welfare of sheep, promoting social responsibility and land management practices without cruelty and exploitation (SCS Global Services, 2023).
- Content Claim Standard (CCS) regulates product content claims and the traceability of specific materials, providing a robust system for tracing raw materials from source to final product (Intertek Total Quality Assured, 2023).
- Global Recycled Standard (GRS) verifies social, environmental, and chemical practices related to the production of products containing recycled materials. Specifically, the standard regulates the production, processing, packaging, labeling, trade, and distribution of all products made with at least 20% recycled material (Textile Exchange, 2023).
- Textile Exchange Organic Content Standard (OCS) verifies the presence of organic material in a product, analyzing the raw materials during harvesting, processing, and production phases (Textile Exchange, 2023).

Specifically in the fashion sector, the main certifications required for sustainability purposes are the following.

Standard 100 by OEKO-TEX (OEKO-TEX Made in Green, 2023) is an independent and uniform international control and certification system for raw materials, semifinished and finished products in the textile sector.

OEKO-TEX's STeP (Sustainable Textile Production) is an independent certification system within the textile supply chain to communicate the sustainability path in a transparent and credible way. Made in Green by OEKO-TEX is an independent textile traceability label. It is applied to consumer products and semifinished products in all production phases of the textile chain, packaged with certified materials, with low environmental impact processes in safe and socially responsible working conditions.

The GOTS—Global Organic Textile Standard (Global Organic Textile Standard, 2023; ICEA, 2023) is an international standard for the issue of an environmental declaration certifying the presence of natural fibers from organic farming in textile products, the maintenance of traceability throughout the production process, restrictions on the use of chemicals and compliance with environmental and social criteria across the textile production chain.

The Organic Content Standard, promoted by Textile Exchange, a nonprofit organization for responsible and sustainable development in the textile sector, provides for the release of an environmental declaration certifying the presence of at least 5% of natural fibers from organic farming and the maintenance of traceability throughout the production process.

NATURTEXTIL IVN (Internationaler Verband der Naturtextilwirtschaft e. V., 2023) is a BEST certified quality standard incorporating guidelines for sustainable fabrics based on definitions of ecology and social responsibility formulated by Internationalen Verband der Naturtextilwirtschaft e. V. (IVN), applied to the entire textile supply chain. The IVN standard sets the highest level of textile sustainability, on the basis of the maximum parameters currently obtainable for the production and the product. IVN BEST certified Naturtextil is a standard for eco-friendly fabrics with the most stringent requirements for ecological textile production at the highest technical levels, both in terms of ecological standards and social responsibility.

GRS is a standard that attributes importance to recycling with the aim of promoting the reduction of resource consumption and increasing the quality of recycled products. The GRS provides for the issue of an environmental declaration that certifies the content of recycled materials, the maintenance of traceability throughout the production process, the restrictions in the use of chemicals, and compliance with environmental and social criteria across all phases of the production chain.

Cradle to Cradle Certified (Product Innovation Institute, 2023) is a globally recognized measure of safer and more sustainable products designed for the circular economy. Cradle to Cradle Certified evaluates products for their environmental and social performance across five sustainability categories: material health, material reuse, renewable energy and carbon management, water management, and social equity. The standard encourages continuous improvement by awarding certification based on five increasing levels of achievement of eco-sustainable requirements.

The GIF (Get it fair) aims to assess the actual risks along the supply chain related to potential dangers for workers, the environment, and local communities. The certification is attributed to companies with a good level of exposure to ESG (Environment, Social, Governance) risks. The brand is issued by the GIF ESG Rating program through an integrated Due Diligence process and the ISO 26000 Guidance for Social Responsibility. The Get It Fair provides consumers with detailed information on the place and working conditions in which a product is made.

The Fairtrade Textile Standard is part of the Fairtrade Textile Program to support change in textile supply chains and resulting business practices to improve working conditions, raise wages, and fair-trading opportunities.

4. Patterns of sustainability in fashion

In pursuing sustainability-based strategy, fashion firms adopt business practices and approaches where environmental and social issues assume distinct values and configurations (Rinaldi, 2020).

Ethical Fashion is based on fairness and pays attention to respect for the environment, ensuring decent working conditions throughout the supply chain (Haug & Busch, 2016; Joergens, 2006; Shen et al., 2012). Traceability and transparency in clothing production are the fundamental elements of the ethical-sustainable approach. The higher costs and investments required to reduce the negative impact on the planet translate into higher prices for final products compared to clothing produced using traditional processes (Choi et al., 2023).

Sustainability transition is evident in *Fast-fashion*, traditionally characterized by "unsustainable" production processes and commercial practices (Vijeyarasa & Liu, 2022). Fast-fashion aims for high sales volumes in a short time and at low prices, resulting in numerous short-lived seasons with negative environmental impact, including material and human resources exploitation, energy consumption, and waste disposal. Ethical and sustainable practices are gradually being introduced that use organic material and promote greater transparency in information relating to the clothing production locations and chemical use, as well as better working conditions in compliance with workers' rights.

Slow-fashion is a counter-trend approach to fast fashion. It is based on low-speed production, use of organic raw materials, small batches production, low consumption of resources and waste reduction, and extended delivery times in compliance with working hours (Domingos et al., 2022). Inspired by limited and high-quality production, organizational well-being, and a slow fashion with an extended clothing life cycle, *Slow-fashion* promotes consumer behavior based on sustainability. It encourages the search for durable clothes, organic materials, enhancement of local resources, local craftsmanship and culture, and a preference for short production chains.

The use of textile waste to create new fibers with low environmental impact is among the frontiers of *Green fashion* in response to fabrics traditionally composed of mixed fibers (natural and synthetic). Biomaterials can have natural origin (such as, e.g., from algae, wood, broom, hemp, coconut, corn, mushrooms, cotton, and bamboo), or synthetic (derived from bio-based polymers). In addition to being biodegradable and recyclable, plant fibers are characterized by ethical practices. Experiments underway among brands, designers, and textile fabric producers are transforming the fashion system (Duarte et al., 2022; Ikram, 2022). Pioneering firms like Calvin Klein, Hugo Boss, and H&M are among those incorporating sustainable practices. An Italian firm, Orange Fiber, is a trailblazer in producing high-quality sustainable fabrics, patenting fibers from citrus waste and by-products from the citrus processing industry (Longo & Faraci, 2021). The design of plants and the production of sustainable fibers are central in the transition to circularity. *Eco-design* applies sustainability principles to the design of products according to the criteria of cost reduction, environmental impacts minimization, and use of renewable energy (Díaz-García et al., 2015). Examples and areas of application of eco-design in the fashion sector include the *zero-waste pattern* to create efficient paper models by limiting the materials waste during the cutting phase; *seamless* clothing; *modular design* to make versatile garments, adding or eliminating parts or accessories of a finished product; and *design for disassembly* to produce clothes with components that can be disassembled based on usage time.

The *second-hand market* emphasizes the logic of reusing products and materials, giving them a second life or another market opportunity. Major brands like H&M, Zara, and Adidas offer collection services, maintenance, repair, and resale of previously used products. *Upcycling* is another practice used to transform one product into another with the same or higher quality through the reuse of some parts of the clothing, while *downcycling* concerns the process of transforming waste into a clothing product having the same or lesser value than the initial one. *Shopping Without Any Payment* (SWAP) is based on the exchange among private individuals of clothes that are no longer used (e.g., Swap Élite, ThredUP).

The circular economy cycle closes with the recycling phase where the waste becomes raw material. Textile waste consists of industrial textile waste obtained from yarns and fabrics; waste from clothing packaging processes; retail returns; and returns from domestic use. Regenerated textile fibers are subsequently used for semifinished products for weaving or knitting.

5. Takeaways for fashion managers: Toward Sustainable Eco-Chic Fashion Tech

The greater awareness of the planet's resources scarsity and the pressing need to preserve the natural heritage's quality lead fashion firms to apply a sustainable approach in production systems and evaluate performance through SDG9. This chapter examined patterns and changes that fashion firms are going through to implement circular economy models. The chapter also highlighted various interventions to favor the achievement of sustainability goals in the fashion sector. Noteworthy among these, are the Charter of Sustainable Fashion for the reduction of greenhouse gas emissions produced by the fashion chain by 2030; EU directives and regulations on the use of textile fibers, labeling and marking of textile products for the traceability of clothes; standards and certifications attesting compliance with eco-sustainability requirements.

Sustainability strategies require a medium—long-term commitment to optimizing the relationship between economic performance and environmental results (Flammer, 2015; Pérez-Bou & Cantista, 2023), and a new business approach (Bastos Rudolph et al., 2023). The Sustainable Eco-Chic Fashion Tech is an innovative approach based on intelligent and sustainable production that is consolidating in the fashion industry. It leverages intelligent technologies and systems based on artificial intelligence to the fashion sector. To ensure a greater efficiency, a waste reduction, and a lower environmental impact, as each production step is optimized for efficiency and data availability (Barendregt & Jaffe, 2020). AI-based systems are being deployed to monitor and improve every aspect of production and supply chains. Such systems have several innovative features including predictive maintenance that uses artificial intelligence to identify problems with equipment and tools in advance to reduce downtime and waste; the energy and water optimization to identify and eliminate inefficiencies in uses; the promotion of efficient production practices using artificial intelligence algorithms. Furthermore, artificial intelligence can be used to design more sustainable products and improve working conditions by enhancing work position monitoring and eliminating risk areas. The Sustainable Eco-Chic Fashion Tech aligns with the eco-friendly fashion trend, catering to consumers seeking more sustainable or low environmental impact clothing (Van Loon et al., 2015). Ecological fashion increases awareness of environmental problems and promotes a more ecological approach to fashion through various actions including: using ecofriendly fabrics; zero-carbon shipping; optimizing waste management through reuse and recycling; adopting ecological packaging systems; using green energy sources; revisiting raw material acquisition processes while protecting ecosystems, fauna and human health. Eco-friendly fashion can be considered a business model but also a social and cultural movement within the fashion industry to promote a radical shift in the way of conceiving and using clothing.

Key takeaways for firms contributing to SDG 9 include:
- Redesigning value chains focusing on traceability and transparency of materials, ethics and circularity.
- Promoting sustainable patterns, including ethical fashion centered on the environment and local communities' respect; slow-fashion promoting responsible consumer actions by focusing on durable clothes, organic materials, and short supply chains; green fashion based on natural fibers and biomaterials use; eco-design characterized by low environmental impact products and waste limitation in design, production, and sales.
- Pursuing objectives of gas emissions reduction and responsible clothing disposal.
- Acting responsibly by proposing sustainable and inclusive collections.
- Renewing infrastructure and production processes with more efficient and environmentally friendly technologies.
- Applying the circular model to the product life cycle to optimize raw materials usage, reduce production waste and inventories, extend product life, and guarantee traceability and transparency in the supply chain.

In conclusion, the chapter suggests reinforcing sustainability by experimenting with digital technologies for green fashion design, emphasizing modularity, versatility, adaptability, and blockchain. Sustainable Eco-Chic Fashion Tech can significantly contribute to pursuing all the three dimensions of sustainability because it combines ecology, efficiency, safeguarding wildlife, and natural ecosystems with fashionable clothing.

References

Avelino, F., & Wittmayer, J. M. (2016). Shifting power relations in sustainability transitions: A multi-actor perspective. *Journal of Environmental Policy and Planning, 18*(5), 628–649.

Barendregt, B., & Jaffe, R. (2020). The paradoxes of eco-chic. In *Green consumption* (pp. 1–16). Routledge.

Bastos Rudolph, L. T., Bassi Suter, M., & Barakat, S. R. (2023). The emergence of a new business approach in the fashion and apparel industry: The ethical retailer. *Journal of Macromarketing, 43*(3), 367–383, 02761467231180456.

CCPB Control and Certification. (2023). Retrieved November 29th 2023 from: https://www.ccpb.it/blog/2017/11/08/global-recycled-standard-grs-tessile-materiali-riciclati/.

Choi, T. M., Feng, L., & Li, Y. (2023). Ethical fashion supply chain operations: Product development and moral hazards. *International Journal of Production Research, 61*(4), 1058—1075.

Díaz-García, C., González-Moreno, Á., & Sáez-Martínez, F. J. (2015). Eco-innovation: Insights from a literature review. *Innovation, 17*(1), 6—23.

de la Motte, H., & Ostlund, A. (2022). Sustainable fashion and textile recycling. *Sustainability, 14*(22), 14903. https://doi.org/10.3390/su142214903

Dey, P. K., & Cheffi, W. (2013). Green supply chain performance measurement using the analytic hierarchy process: A comparative analysis of manufacturing organisations. *Production Planning and Control, 24*(8—9), 702—720.

Domingos, M., Vale, V. T., & Faria, S. (2022). Slow fashion consumer behavior: A literature review. *Sustainability, 14*(5), 2860.

Duarte, L. O., Vasques, R. A., Fonseca Filho, H., Baruque-Ramos, J., & Nakano, D. (2022). From fashion to farm: Green marketing innovation strategies in the Brazilian organic cotton ecosystem. *Journal of Cleaner Production, 360*, 132196.

Dwivedi, A., Agrawal, D., & Madaan, J. (2019). Sustainable manufacturing evaluation model focusing leather industries in India: A TISM approach. *Journal of Science and Technology Policy Management, 10*(2), 319—359.

European Commission. (2022). *Corporate sustainability due diligence. Fostering sustainability in corporate governance and management systems.* https://commission.europa.eu/business-economy-euro/doing-business-eu/corporate-sustainability-due-diligence_en.

European Commission. (2009). *Ecodesign directive 2009/125/EC.* https://eur-lex.europa.eu/legal-content/EN/TXT/PDF/?uri=CELEX:02009L0125-20121204&from=EN.

European Parliament and of the Council. (2022). *Directive (EU) 2022/2464 of the European Parliament and of the Council of 14 December 2022 amending regulation (EU) No 537/2014, directive 2004/109/EC, directive 2006/43/EC and directive 2013/34/EU, as regards corporate sustainability reporting (text with EEA relevance).* https://eur-lex.europa.eu/legal-content/EN/TXT/?uri=CELEX:32022L2464.

European Parliament and of the Council. (2006). *Regulation (EC) n. 1907/2006 concerning the registration, evaluation, authorization and restriction of chemicals (REACH), establishing a European Chemicals Agency, amending directive 1999/45/EC and repealing regulation (EEC) no. 793/93 of the council and regulation (EC) no. 1488/94 of the commission, as well as council directive 76/769/EEC and commission directives 91/155/EEC, 93/67/EEC, 93/105/EC and 2000/21/EC.* https://eur-lex.europa.eu/legal-content/EN/TXT/?uri=CELEX%3A0 2006R1907-20140410.

European Parliament and of the Council. (2004). *Regulation (EC) no. 850/2004 on persistent organic pollutants and amending directive 79/117/EEC.* https://faolex.fao.org/docs/pdf/eur87038.pdf.

European Parliament and of the Council. (2011). *Regulation (EU) no. 1007/2011 on the names of textile fibers and on the labeling and marking of the fibrous composition of textile products and which repeals directive 73/44/EEC of the council and the directives of the European Parliament and Council 96/73/EC and 2008/121/EC.* https://eur-lex.europa.eu/legal-content/EN/TXT/?uri=celex%3A32011R1007.

European Parliament and of the Council. (2012). *Regulation (EU) no 528/2012 concerning the making available on the market and use of biocidal products text with EEA relevance.* https://eur-lex.europa.eu/legal-content/EN/TXT/?uri=celex%3A32012R0528.

Flammer, C. (2015). Does corporate social responsibility lead to superior financial performance? A regression discontinuity approach. *Management Science, 61*(11), 2549—2568.

Global Fashion Group. (2020). *Sustainability fashion report, people, planet positive. Worldwide.* https://global-fashion-group.com/wp-content/uploads/2021/03/PPP.pdf.

Global Organic Textile Standard. (2023). *Ecology & social responsibility.* Retrieved November 29th 2023 from: https://www.global-standard.org/.

Haug, A., & Busch, J. (2016). Towards an ethical fashion framework. *Fashion Theory, 20*(3), 317–339.

ICEA. (2023). *Global Organic textile standard*. Retrieved November 29th 2023, from: https://icea.bio/certificazioni/non-food/prodotti-tessili-biologici-e-sostenibili/global-organic-textile-standard/.

Ikram, M. (2022). Transition toward green economy: Technological Innovation's role in the fashion industry. *Current Opinion in Green and Sustainable Chemistry*, 100657.

Internationaler Verband der Naturtextilwirtschaft e. V. (2023). Retrieved November 29th 2023 from: https://naturtextil.de/en/ivn-quality-seals/about-naturtextil-ivn-zertifiziert-best/.

Intertek Total Quality Assured. (2023). Retrieved November 29th 2023 from: https://www.intertek.com/assurance/ccs/.

Joergens, C. (2006). Ethical fashion: Myth or future trend? *Journal of Fashion Marketing and Management: International Journal, 10*(3), 360–371.

Jovanovich, N. (2022). *What is eco-friendly fashion and why it is important*. https://ecobnb.com/blog/2022/12/eco-friendly-fashion-important/.

Klassen, R. D., & Johnson, P. F. (2004). The green supply chain. Understanding supply chains: Concepts, critiques and futures. *Green Supply Chain*, 229–251.

Longo, M. C., & Faraci, R. (2021). Lo storytelling nelle imprese equity-based. Il Caso Orange Fiber, da start-up a PMI innovativa di successo. In *Quaderni di ricerca sull'artigianato, Rivista di Economia, Cultura e Ricerca Sociale" 1/2021* (pp. 79–112).

Loorbach, D., & Wijsman, K. (2013). Business transition management: Exploring a new role for business in sustainability transitions. *Journal of Cleaner Production, 45*, 20–28.

Markard, J., Raven, R., & Truffer, B. (2012). Sustainability transitions: An emerging field of research and its prospects. *Research Policy, 41*(6), 955–967.

Martel, A., & Klibi, W. (2016). *Designing value-creating supply chain networks*. Cham: Springer.

Mukendi, A., Davies, I., Glozer, S., & McDonagh, P. (2020). Sustainable fashion: Current and future research directions. *European Journal of Marketing, 54*(11), 2873–2909. https://doi.org/10.1108/EJM-02-2019-0132

NSF. (2023). Retrieved November 29th 2023 from: https://www.nsf.org/knowledge-library/sustainable-down-certifications.

OECD. (2001). *Extended producer responsibility: A guidance manual for governments*. Paris: OECD Publications Service. https://www.oecd.org/environment/extended-producer-responsibility.htm.

OEKO-TEX Made in Green. (2023). Retrieved November 29th 2023 from: https://www.oeko-tex.com/en/our-standards/made-in-green-by-oeko-tex.

Pérez-Bou, S., & Cantista, I. (2023). Politics, sustainability and innovation in fast fashion and luxury fashion groups. *International Journal of Fashion Design, Technology and Education, 16*(1), 46–56.

Papamichael, I., Chatziparaskeva, G., Pedreno, J. N., Voukkali, I., Candel, M. B. A., & Zorpas, A. A. (2022). Building a new mind set in tomorrow fashion development through circular strategy models in the framework of waste management. *Current Opinion in Green and Sustainable Chemistry, 36*, 100638.

Product Innovation Institute, 2023 Retrieved November 29th 2023 from:https://www.c2ccertified.org/.

Rinaldi, F. R. (2020). *Fashion industry 2030: Reshaping the future through sustainability and responsible innovation*. Milano: EGEA.

Saxena, R., & Khandewal, P. K. (2010). Sustainable development through green marketing: The industry perspective. *The International Journal of Environment, Cultural, Economic and Social Sustainability, 6*(6), 59–79.

SCS Global Services. (2023). Retrieved November 29th 2023 from: https://www.scsglobalservices.com/services/responsible-wool-standard.

Shen, B., Wang, Y., Lo, C. K., & Shum, M. (2012). The impact of ethical fashion on consumer purchase behavior. *Journal of Fashion Marketing and Management: International Journal, 16*(2), 234–245.

Sinha, P., Sharma, M., & Agrawal, R. (2023). A systematic review and future research agenda for sustainable fashion in the apparel industry, Benchmarking. *International Journal, 30*(9), 3482–3507. https://doi.org/10.1108/BIJ-02-2022-0142

Textile Exchange. (2023). Retrieved November 29th 2023 from: https://textileexchange.org/app/uploads/2021/02/GRS-v4.2-Implementation-Manual.pdf.

Textile Exchange. (2023). *Textile Exchange Organic Content Standard*. Retrieved November 29th 2023 from: https://textileexchange.org/.

Thakker, A. M., & Sun, D. (2023). Sustainable development goals for textiles and fashion. *Environmental Science and Pollution Research, 30*(46), 101989–102009.

Thorisdottir, T. S., & Johannsdottir, L. (2019). Sustainability within fashion business models: A systematic literature review. *Sustainability, 11*(8), 2233.

UNECE. (2023). *SDG 9 & fashion – industry, innovation and infrastructure*. Retrieved November 29th 2023 from: https://unece.org/forests/events/sdg-9-fashion-industry-innovation-and-infrastructure.

Van Loon, P., Deketele, L., Dewaele, J., McKinnon, A., & Rutherford, C. (2015). A comparative analysis of carbon emissions from online retailing of fast moving consumer goods. *Journal of Cleaner Production, 106*, 478–486.

Vijeyarasa, R., & Liu, M. (2022). Fast fashion for 2030: Using the pattern of the sustainable development goals (SDGs) to cut a more gender-just fashion sector. *Business and Human Rights Journal, 7*(1), 45–66.

Yazdanifard, R., & Mercy, I. E. (2011). The impact of green marketing on customer satisfaction and environmental safety. *International Conference on Computer Communication and Management, 5*, 637–641.

Zimmer, K., Fröhling, M., & Schultmann, F. (2016). Sustainable supplier management—a review of models supporting sustainable supplier selection, monitoring and development. *International Journal of Production Research, 54*(5), 1412–1442.

CHAPTER SIX

Ensuring sustainable patterns in tourism

1. SDG 12 for sustainable tourism

SDG 12—Responsible consumption and production aims to adopt sustainable consumption and production (SCP) methods, supporting the path toward sustainability. It promotes the implementation of the UN's 10-year program for an environmentally friendly approach to chemical use and waste management. The firms and organizations involved in achieving the objective are called to manage, in a sustainable way, resources, energy, and infrastructures, to adopt sustainability criteria in the purchases of goods and services, and to guarantee dignified jobs and organizational well-being. Sustainable consumption and production goal aims to improve the quality of products with a more efficient use of resources. Achieving Goal 12 requires international and national regulation and integrated sector plans to responsibly manage resources, implement sustainable business practices, and raise awareness of consumer behavior. This implies an increase in economic and social benefits with a reduction in the use of resources, pollution, and waste along the production cycle. The SDG 12 targets are articulated in various areas: implementation of the 10-year sustainable consumption and production framework; sustainable management and use of natural resources; halving global per capita food waste; responsible management of chemicals and waste; substantial reduction of waste production; adoption of sustainable practices and sustainability reporting; promotion of sustainable public procurement practices; universal understanding of sustainable lifestyles; supporting developing countries' scientific and technological capacity for sustainable consumption and production; developing and implementing tools to monitor sustainable tourism; removing market distortions that encourage wasteful consumption.

This objective specifically mentions tourism as a sector of intervention relevant and cross-cutting for its impact on energy and water consumption, the production and disposal of waste, the protection of biodiversity, employment, the promotion of local culture and products, and the quality

Being a Sustainable Firm
ISBN: 978-0-443-14062-4
https://doi.org/10.1016/B978-0-443-14062-4.00006-4

of life in tourist destinations (Pjerotic et al., 2017; Lenzen et al., 2018). Considering these aspects, recognizing tourism as a key sector for SDG 12 in the commitment to implement SCP practices means attributing to it the central role of accelerator of the global transition toward sustainability.

Sustainable development is the central concept of development and tourism has enormous economic, environmental, and social impact in the modern world; therefore, the development of this sector is linked with all three mentioned dimensions of sustainability (Khan et al., 2020). Tourism represents a social, cultural, and economic phenomenon that has an impact on the economy, nature, and local communities. It is connected to the movement of people to different destinations, responding to the needs of visitors and stakeholders in host communities. Sustainable tourism is a multidimensional concept that includes several areas, such as the maintenance of ecology, the conservation of biodiversity, and the optimal use of natural resources; the respect for host communities protecting their cultural heritage and traditions; a firm management that takes into consideration all stakeholders' interests and social opportunities for the community (UNWTO World Tourism Organization, 2023). Sustainable tourism is linked to social responsibility as part of its implementation. Consistent with environmental, economic, and social objectives, sustainable tourism seeks to balance the need to satisfy tourists through meaningful experiences and raise awareness of sustainability issues. Table 6.1 reports the categories of consumption products related to the tourism industry (United Nations, 2010).

Sustainable tourism is based on the assumptions of protecting environmental, economic, social, and cultural resources to contribute to the preservation of tourist destinations and surrounding territories. Studies on sustainable tourism have highlighted transition processes toward tourism sustainability (Gössling et al., 2012; Köhler et al., 2019; Loorbach & Wijsman, 2013 ; Markard et al., 2012), the role of digital technologies in transforming tourism services (Bilgihan & Nejad, 2015; Ali et al., 2020; Kasemsap, 2017), the design of sustainable business models for new businesses (Coles et al., 2017; Szromek & Herman, 2019), organizational restructuring, renewal and governance strategies (Díaz-García et al., 2015; Mantaguti & Mingotto, 2016), the role and involvement of the local community (Lee, 2013; Lim & McAleer, 2005) and the implications of sustainability choices in different entrepreneurial contexts (Bynum & Muzaffer, 2013; Díaz-García et al., 2015; Hassan, 2000; Romagosa, 2020;

Table 6.1 Categories of tourism characteristic consumption products and activities (tourism industries).

Consumption products	Activities/industries
1. Accommodation services for visitors	1. Accommodation for visitors
2. Food and beverage serving services	2. Food and beverage serving activities
3. Railway passenger transport services	3. Railway passenger transport
4. Road passenger transport services	4. Road passenger transport
5. Water passenger transport services	5. Water passenger transport
6. Air passenger transport services	6. Air passenger transport
7. Transport equipment rental services	7. Transport equipment rental
8. Travel agencies and other reservation services	8. Travel agencies and other reservation services activities
9. Cultural services	9. Cultural activities
10. Sports and recreational services	10. Sports and recreational activities
11. Country-specific tourism characteristic goods	11. Retail trade of country-specific tourism characteristic goods
12. Country-specific tourism characteristic services	12. Other country-specific tourism characteristic activities

From United Nations (2010).

Sharma et al., 2020). To the SDG 12 achievement, the importance of resource co-management is increasingly consolidating through participatory decision-making processes and sharing of environmental, economic, and social sustainability principles with local communities (Prayag et al., 2010; Rahman et al., 2021). A further theme, within the theoretical debate on sustainable tourismconcerns the ability of tourism businesses to attract capital and effectively measure the impact of sustainability strategies through reporting systems and sustainability performance indicators (Jones et al., 2016; Prud'homme & Raymond, 2016; Dwivedi et al., 2019). The different perspectives of sustainable development should be reconciled with the competitiveness needs of the sector. Complex ecological problems concern waste management and sorting, deseasonalization, energy saving, product recycling, cultural and environmental resources protection, and soil, sky, and water pollution. Another critical issue concerns the definition of reliable indicators to measure the progress of sustainable development in tourism and the effectiveness of sustainability policies (Madhavan & Rastogi, 2013; Agyeiwaah et al., 2017; Waseema, 2017). Indicators and evaluation methods consider the relationship between economic activities and implications on socio-environmental performance.

The United Nations Ten-Year Framework on Sustainable Consumption and Production Patterns (UN 10-Year Framework of Programs) adopts a specific Sustainable Tourism Program (STP) under the direction of the World Tourism Organization (UNWTO). It promotes partnerships for the SDG12 implementation and related objectives, including climate actions and the protection of marine and terrestrial ecosystems, along with a series of projects focusing on criteria, certifications, and standards for designating sustainable tourism destinations and responsible services. The program increases awareness of the real and positive impacts of tourism by influencing public decisions and business choices. It also promotes initiatives to encourage sustainable lifestyles, through the dissemination of information on labels and compliance with standards and certifications.

There are various stakeholders involved in SDG 12 achievement, including firms, consumers, political decision makers, universities and research centers, distribution channels, media, and institutions. The One Planet network was established to support the SDG 12 implementation through a multilateral partnership for sustainable development. The network provides operational tools and solutions for accelerating the transition toward sustainable consumption and production (www.oneplanetnetwork.org).

The Global Destination Sustainability Index (GDS-Index) develops one of the most important rankings worldwide on the degree of improvement in performance that makes business and leisure tourism more sustainable (https://www.gds.earth/index/). This indicator evaluates four key areas of a destination's sustainability performance, such as the city's environmental and infrastructure strategy, the city's social sustainability performance, industry supplier support, and the organization's strategy and initiatives of destination management. The 2023 GDS-Index shows the ranking of global destinations for sustainability. First place, for the seventh time, is occupied by Gothenburg (Sweden) with a score of 94.64%. This city, named the European Capital of Smart Tourism in 2020, also earned the title of World's Best Sustainable City Stay 2021 by Lonely Planet. Oslo, Copenhagen, and Helsinki follow in the GDS-Index ranking. Emerging trends include climate strategies, stakeholder engagement, certification, social impact actions, and initiatives on diversity, equity, and inclusion. The GDS-Index trend shows that destination management organizations and national tourism entities are the main drivers of economic, social, and environmental transformation due to their activity within tourism ecosystems and regenerative action plans. Central points of Gothenburg's SDG 12 strategy are programs relating to design, accessibility, and broad collaboration between administrations and firms. In 2018 Wonderful Copenhagen focused its sustainability strategy on the theme of "Tourism for Good" to guarantee a positive impact of tourism on the territory. In 2019, the city received the Green Tourism Organization certification, for the support provided to destinations and organizations involved in sustainability. Lastly, the "European Capital of Intelligent Tourism 2019" was awarded by the European Union to Lyon for its achievements in digital innovations.

The Meaningful Tourism Index (www.meaningful-tourism.com) is another important index to measure the positive effects of sustainability. It starts from the assumption that sustainable tourism development can be achieved with the participation of six categories of stakeholders that are travelers, host communities, staff, firms supplying tourism and accommodation services, governments at different levels, and the environment (i.e., future generations).

The Meaningful Tourism Index 2023 measures these effects through 72 indicators, of which 12 refer to the overall situation of tourism and the tourism industry in a country and 10 indicators individually concern the six main stakeholder groups. The index includes all UN member countries that recorded more than two billion dollars in international tourism receipts in one of the previous 5 years (2018–22), for a total of 88 countries. The index adopts a gradual approach divided into three steps: analysis of the current situation in terms of impact on the quality, strengths and weaknesses, benefits, and satisfaction of tourism for each of the six stakeholders; definition of key performance indicators and measurement methods for stakeholders including them in the decision-making process; identification of contradictions and methods to align the different interests. The requirements of each stakeholder are considered key performance indicators (KPIs). The analysis of contradictions is based on the search for convergent solutions among stakeholders. They are not viewed as competitors in conflict with each other and with prevailing and opposing interests; instead, they are considered interested parties who collaborate by focusing on common objectives to ensure the achievement of positive KPIs for all parties.

2. Key enablers of sustainable digital tourism

Digital tourism is characterized by the application of digital technologies to the tourism industry (Adeola & Evans, 2019). It can be defined as an interdisciplinary field in which elements of virtual tourism interact with the customer experience, changing the way of doing business, living, and traveling. The keywords associated with digital tourism include sustainability, smart tourism, e-tourism, social media, destination marketing, big data, smart cities, smart tourism destination, destination image, and satisfaction (Kalia et al., 2022). Digital tourism makes activities more attractive and productive across all travel phases, generating new business opportunities and innovative ways of organizing and offering tourist services to improve the customer experience. The key enablers of digital transformation encompass various areas and actors within the tourism ecosystem, such as the business processes organization, online booking and sales system, image dissemination and destination brand creation, online reviews and experience evaluation, proposals and purchases on digital platforms, and metaverse experience (Longo & Faraci, 2023).

Traditionally, digital tourism has been associated with e-commerce and digital marketing; however, the concept is much broader as it concerns the direct and continuous connectivity between customers and firms in the tourism system. The primary tools for improving sustainable digital tourism are innovation, mobile devices, and technologies (Li et al., 2020), social media and inclusion (Mehraliyev et al., 2019), smart tourist destination (Bastidas-Manzano et al., 2021), sustainability and competitiveness (Seguì-Amortegui et al., 2019). It is precisely in the promotion of sustainable tourism that digital technology emerges as a central factor (Della Corte et al., 2019; Li et al., 2022) in enhancing the destination and the tourist experience. This amplifies the impact of sustainability strategies on tourist destinations and improves the management of tourism resources by reducing environmental pollution and creating social value for local residents and tourists in the destination itself. Another area of focus concerns the possibility of increasing the communication effectiveness, reaching increasingly specific targets, and finding a balance between tourists' expectations and the needs of host communities, thereby protecting natural and cultural resources (Lozano-Oyola et al., 2012; Purwanda & Achmad, 2022). The Tourism Digital Hub is a web platform, based on artificial intelligence models, designed to integrate the tourism ecosystem represented by tourism operators, firms from different sectors, and institutional stakeholders. The main objective is to digitally connect the offering of tourist products, develop an information system for the promotion of tourist activities and territories, expand the portfolio of services offered, and assist tourists in planning trips and organizing tours. Big data from different sources (Member States, sectors, businesses, local and regional authorities, universities, institutions) flows into the Common European Tourism data space (Directorate-General for Internal Market, Industry, Entrepreneurship and SMEs, 2023). This organization defines a governance framework for the common space in compliance with national and European legislation and EU standards. Data and information, such as those relating to the energy consumption of tourist facilities, waste and recycling, and food consumption, can be used to monitor the environmental impact of tourism in a specific area and to take corrective actions. This system ensures interoperability among key sectoral data spaces for energy, mobility, environment, health, well-being, smart community, cultural heritage, and sectors related to the tourism experience. The Common European tourism data space allows users, tourism intermediaries, providers of tourist services and packages, and actors directly or indirectly interested in tourist flows to access data. Firms and public institutions can use data and information to develop innovative tailored-based services and support strategic choices related to tourism sustainability. This activity is supported by various programs, including the European Tourism Agenda 2030, which provides for the implementation of data-sharing practices on tourism and ecotourism, and Digital Europe (DEP), which has developed the EU Tourism Dashboard to monitor green and digital transitions and the resilience of the tourism ecosystem.

Digitalization concerns tools for supporting promotion to attract tourists in the decision-making process and tools for facilitating the use of services and local resources, supporting tourists during their visit (Gretzel et al., 2018). The integration between ICT and innovative digital tools, connected to the web and the cloud system, favors targeted customization and simplifies the search for information. Some technologies that contribute to create value also in terms of sustainable tourism are the following. The Geographic Information System (GIS) is a system to simplify procedures and make access to information through geospatial data easier. It reproduces thematic maps integrating them with descriptive information and position data. The Yield Management systems allow airlines and hotel companies to adjust rates based on demand trends (Viglia & Abrate, 2020; Gabor et al., 2022). This system optimizes, in terms of economic sustainability, the levels of revenue and profitability of firms, while simultaneously offering competitive rates for customers. Electronic travel agencies (such as Expedia, Booking, Trivago) favors the direct sales of airlines and accommodation facilities. Web tools simplify relationships and content sharing, while mobile technologies reduce the space-time boundaries between firms and users, providing real-time information and updates. Software for Customer Relationship Management Systems (CRM) collects data related to individual visitors to personalize offers and establish a long-term relationship. Finally, e-marketing tools amplify the range of communication channels, making information and promotional campaigns more efficient.

3. Corporate social responsibility and thematic areas in sustainable tourism

Corporate social responsibility (CSR) concerns the behavior and responsibility of firms regarding their environmental and social impact on the economy and the territorial context in which they operate (Hatipoglu et al., 2019; James, 2012; McWilliams & Siegel, 2001; Uyar et al., 2023). The European Commission defines CSR as "a concept whereby companies integrate social and environmental concerns in their business operations and in their interaction with their stakeholders on a voluntary basis" (Commission of the European Communities, COM (2006) 136 final, 2006). Stakeholders have social expectations that go beyond mere compliance with the law, focusing on the ethical and moral aspects of business conduct (Carroll, 2016). Compliance with implicit and explicit rules for the protection and respect of all interested parties and the adoption of minimum legal requirements in the supply of goods and services are prerequisites of the social and formal responsibility of firms. In addition to these rules, there are ethical expectations more internal and immaterial than laws and procedural regulations. CSR includes respect for ethical and moral principles and recognition of the values of integrity and behavioral ethics (Bolton & Benn, 2010; Madanaguli et al., 2022), and requires that all initiatives are voluntary.

CSR in tourism has great potential in preserving the environment and creating an inclusive and equitable business environment. In a CSR approach, tourism businesses face several challenges in reconciling profit-making purposes with obligations and responsibility toward societies. Relevant issues in the connection between businesses, environment, and society are as follows: the seasonality of tourism, which leads to precariousness, low wages, and long work shifts; the development of skills associated with increased levels of employment and decent work; the contribution to the reduction of poverty and the economic growth of the territory, especially if underdeveloped; the protection of natural ecosystems; and the interculturality of tourist destinations. Businesses can address these issues with responsible strategies and policies, where sustainability, ethics, morality, society, and the environment are central elements of CSR.

The CSR benefits for firms are manifolds. The first concerns the achievement of positive economic and environmental outcomes through more efficient use of natural resources, energy-saving management, and specific waste recycling and reuse programs (González-Morales et al., 2016; Levy & Park, 2011). The second advantage derives from CRS practices generating relevant data and informationfor stakeholders on social and environmental actions and policies adopted. The third benefit refers to the development of standardized methodologies and shared reporting systems (Ettinger et al., 2018; Rodrigues & Mendes, 2018). Finally, CSR focuses on evaluating the degree of satisfaction and perception of employees and customers with respect to the responsible approach adopted by the firm (Kucukusta et al., 2013; Tsai et al., 2012). In this way, both employees and customers of a tourism firm can perceive the value of the organization's commitment to the actual environmental and local community policies. To this end, the implementation of CSR programs should be consistent with the firm's mission and vision for improving services and corporate image (Serra et al., 2018).

In this context, actions and tools consistent with CSR objectives are as follows: the use of green transport and the selection of the travel distances based on ecological criteria, information provided to customers on the CO_2 emissions of the offered package, availability of accommodation facilities that meet environmental and social standards; destination proposals that prioritize sustainability; respect for local communities and intercultural integration; application of appropriate pricing that covers all costs while also guaranteeing decent wages; compliance with workplace safety regulations and support for disadvantaged and weak categories; protection of children from forms of exploitation; transparency in partnerships, networks, and relationships with its stakeholders. CSR-oriented tourism businesses can contribute to the protection of sensitive areas such as coral reefs, nature reserves, and coastal areas, reducing waste and pollution in hotels, resorts, and tourist villages. The establishment of tourism activities should avoid damaging local habitats and maintain the quality of life in the destination. Furthermore, investments in destination development should take into account the attractiveness of the areas influenced by rising sea levels and risks of floods, storms, and hurricanes. Forms of experiential tourism linked to winter sports should consider the risk of avalanches and landslides caused by rising temperatures. Hospitality companies should be prepared for security issues and implement coordination procedures with the government to protect guests in emergency situations arising from acts of terrorism, wars, and civil unrest, health crises, and natural disasters. When it comes to travel, it is important to provide passenger reprotection measures, follow a management approach for risk information and provide effective solutions in emergency situations. The critical Environmental and Social Governance (ESG) issues identification impacting the tourism business is subject to continuous evaluation to have greater harmonization between common risks, goals, and answers.

The World Travel and Tourism Council (WTTC) monitors sustainability reporting in the tourism sector. After consulting with experts and stakeholders in the field of sustainability, the organization has identified several thematic areas representing issues and evolutionary lines for a balanced and sustainable future of tourism. The main themes (World Travel & Tourism Council, 2017) outlined refer to:
- preserving the sustainability of destinations in an evolving world;
- adopting responsible business practices and promoting innovation in travel and tourism activities;
- encouraging changes in the labor market and employment practices;
- associating travel and forms of tourism with the environment, health, human rights and security;
- taking action on issues relating to climate change, destination degradation, and poverty.

The aforementioned thematic areas of interest for tourism are represented in the sustainability reporting systems. It is essential to define appropriate models for the detection, communication, and representation of social and environmental variables concerning tourism thematic areas to have a comprehensive framework of sustainability.

Therefore, CSR is linked to the ESG approach, as both pursue the sustainability of society and the well-being of the planet. However, the two approaches differ in methods and measurement of results. CSR is a form of self-regulation with a philanthropic character. It proposes a general framework of sustainability that becomes part of the culture of an organization. It anticipated ESG issues, laying the foundation for the development of a responsible firm. The ESG approach defines a methodological framework

Table 6.2 Comparing CSR and ESG.

Corporate social responsibility	Environmental and social governance
Qualitative approach	Qualitative and quantitative approach
Self-regulated	Regulated by rules
Based on principles, values, and philosophy of responsibility	Based on measurements, standards, evaluation criteria
It feeds responsibility and accountability	It documents responsibility objectively

Authors' elaboration.

for evaluating long-term sustainability performance by combining the three dimensions. Recently, the concept has been extended to quantifiable and/or objectively documentable corporate sustainability practices. ESG measurement criteria allow the definition of documentable sustainability indicators to measure responsibility. Based on the data, rating agencies evaluate ESG performance. Table 6.2 highlights the main features characterizing CSR and ESG approaches.

4. Sustainability reporting and disclosure information in tourism

The definition of standards in the sustainable tourism sector is connected to the need of increasing the quality of sustainability reporting, which is often incomplete and reliable. A clear vision of corporate risks related to sustainability requires firms to provide information on their strategies and plans for mitigating negative impacts on the environment, green investments, and all activities related to environmental and community protection. In 2023, the European Commission officially adopted the European Sustainability Reporting Standards (ESRS) developed by the European Financial Reporting Advisory Group (EFRAG), as relevant guidelines for tourism within the European territory. EFRAG, as a technical advisor, provides its expertise and role to support firms in sustainable reporting, answering questions on the legal interpretation of ESRS standards and alignment with the standards issued by the International Sustainability Standards Board (ISSB) and with the Global Reporting Initiative (GRI).

The standards, which are expected to come into force in January 2024, are consistent with the Corporate Sustainability Reporting Directive (CSRD, 2022) and applicable to all companies operating in the tourism sector. Their purpose is to require large corporations and all listed companies to evaluate and communicate their actions, risks, impacts, and opportunities with respect to social and environmental issues. The ESRS standards are, therefore, mandatory for firms already obliged by CSRD to report and communicate specific information on sustainability. These reporting systems should be complied with the criteria and procedures established by the CSRDs.

The adoption of these standards requires the reporting of comparable and reliable information on sustainability, aiming to assist firms in reducing reporting costs and avoiding voluntary and nonuniform approaches in reporting and communication. This information represents a basis available to managers, investors, social organizations, consumers, and institutions to gain awareness of activity management, evaluate the sustainable performance of tourism businesses, and facilitate access to sustainable finance. Among the general requirements, the ESRS 1 standard outlines the prerequisites that characterize effective and complete sustainability reporting. Specifically, essential elements to monitor include the ways in which resources are conserved, the commitment to implementing responsible tourism practices, and the local community involvement. The ESRS 2 standard on general disclosure underlines the importance of transparency in the communication and dissemination of information on sustainability practices, with particular reference to eco-compatible policies and initiatives, cultural conservation actions, and tourism projects activating participatory processes with the community.

Following the entry into force of the European Directive on Corporate Sustainability Reporting (CSRD) in 2023, the ESG approach becomes increasingly relevant due to the mandatory ESG reporting starting in 2025. The report, based on specific standards, provides a way to measure sustainability performance, including risks and opportunities in the three ESG areas. Financial performance is critical to the ESG rating, because as a company's ESG score increases, its capital costs reduce, improving the company's overall rating. The main purpose of ESG is to objectively document the responsibility and implementation of systems and processes to measure environmental impact. Therefore, ESG detects relevant non financial information and data that are not contained in traditional accounting systems.

The tourism industry is oriented toward ESG reporting on sustainability also following various regulatory interventions and requests from customers and institutions. The tourism ESG report captures the interactions among tourism operators, the environment, and the community and evaluates non financial areas such as customer acquisition and retention, risk management, investor relations, and access to sustainable financing. The co-presence of different sizes of enterprises and the variety of tourism activities produce a high degree of heterogeneity in reporting models. Additionally, as firms operate in different markets, there is a need to address issues and impacts of global scope and context-related. The World Travel and Tourism Council research (2017) analyzed the adoption of sustainability reporting in five tourism categories (airlines, hotel or branded company operators, cruise lines, travel agencies/operators and global distribution systems (GDS), and technology providers). An organization is considered a "reporter" if it has produced, within a specific time frame, at least one sustainability report or similar (environmental balance sheet, social balance sheet) of 10 pages with quantitative data. Research results showed a 50% increase in reporters in the Travel and Tourism sector, despite the number of potential reporters growing by 4%. Within the categories, reporting by airlines has grown more than that by travel agencies and tour operators. Several organizations have reported sustainability using specific indicators to report information, without using the GRI methodology or including the materiality matrix of key topics ordered by degree of impact on business and society. As a result, the current status of the tourism sector is characterized by a wide range of indicators and measurements that overlap or are not comparable.

The material topics reported by the Global Reporting Initiative (GRI) in the tourism sector are as follows: safety practices, public order, labor/management relations, compliance, customer privacy, customer health and safety, no discrimination, cleanliness, indigenous rights, and child labor. According to the Global Reporting Initiatives Tourism 2030, the 10 actions to achieve the SDGs2030 are those listed below.

- *Natural heritage and biodiversity.* It refers to the biodiversity of ecosystems, the species present on the planet, and the value of natural assets.
- *Human rights and labor rights.* It is about improving the well-being of local people and the dignity of workers, with the aim of reducing exploitation, inequality, and poverty.
- *Good governance and CSR.* It concerns participation, transparency, and responsibility in the implementation of good governance processes of tourist destinations at all administrative levels.
- *Travel, transport, and mobility.* It refers to the use of transport means with low environmental impact ("soft mobility" or "sustainable mobility") to reduce the ecological footprint of travelers.
- *Destination management.* It involves coordinated partnership actions following the principles of good governance to implement a territorial approach to the multisectoral, multilateral, and multithematic matrix of tourism development.

- *Knowledge, networking training, and education.* It consists of investing in human potential through networking and education to promote innovation and address the common challenges of sustainable development.
- *Climate change—energy and resource efficiency.* It refers to innovative management of resource efficiency following climate change and the need for adaptation to the resulting impacts.
- *Cultural heritage, lifestyles, and diversity.* It refers to the protection of cultural heritage in its material and immaterial forms, including traditions, lifestyles, and customs, as an integral part of the drive toward sustainable development.
- *Value chain management and fair trade.* It concerns the management of the tourism value chain and the diffusion of fair trade for responsible tourism. The creation of global green value chains can encourage the development of fair-trade practices.
- *Certification and marketing.* Certification and labeling certify the presence of sustainable tourism products in a transparent and understandable way, helping travelers and business partners to choose responsibly.

Another tourism sustainability measurement system is the Statistical Framework for Measuring Tourism Sustainability (SF-MST). It is designed to collect data on the economic situation and on environmental and social effects holistically and on a geographical scale. Themes and potential indicators are sistematically grouped and classified based on the degree of the economic system's connection with environmental and social dimensions. The coherence among different indicators allows us to evaluate the validity of sustainability choices and policies and to compare different destinations. The set of key indicators for tourism sustainability reporting by SF-MST are classified by dimension (general indicators, economic, environmental, social), measurement theme (such as visitor length of stay, tourism concentration, tourism visitor dependency, seasonality, visitor expenditure, tourism economic structure and performance, employment, investment and government tourism related-transactions, solid waste flows, water and energy flows, ecosystem services flows, visitor satisfaction, host community perception, decent work and governance) and potential indicators (such as average length of stay, number of visitors and variations, average internal tourism expediture per visitor, tourism direct GDP, totale employment and labor productivity, number of repeated visitors, overall perception of host communities of visitors, share of compensation of employed persons, and implementation of standards accounting tools.

5. Takeaways for tourism managers: Moving forward on regenerative tourism

There is a general consensus on the significant contribution of tourism to the economic, environmental, and social growth of a destination and the community that lives there (Asmelash & Kumar, 2019; Brida et al., 2016; Lean, 2009). The negative impacts that tourism can generate on the environment, landscapes, and society are also undisputed. The SDG12 objective on sustainable production and consumption models poses significant challenges for sustainable tourism (Uysal et al., 2016). Among these, the fight against climate change, food security, human health and well-being, quality of life, disasters and major crisis management, and over and under tourism represent key topics that involve businesses and workers, residents, travelers, and visitors. Responding to these challenges requires changing current production and consumption models toward sustainable forms to achieve efficient management of natural resources, processes to reduce food waste and overall waste, and eco-sustainable use of chemical products.

The concept of tourism and the way of doing and experiencing tourism are transforming (Bramwell et al., 2017; Higham & Miller, 2018; Pung et al., 2020; Pung et al., 2022) to the point of taking on the characteristics of regenerative tourism. Regenerative tourism represents a sustainable way of traveling and experiencing places. It goes beyond the notion of visiting without damaging the environment, aiming instead to revitalize the surroundings by activating a positive cycle of impacts on communities and local economies. It denotes the ability of tourism activities to positively contribute to human and non human well-being, giving something back to the host destination, and regenerating it (Becken & Kaur, 2021). From this perspective, regenerative tourism evaluates the cost–benefit and value ratios it generates with respect to stakeholders and living beings that populate a territory (Bellato et al., 2022; Cave & Dredge, 2020; Cave et al., 2022). The concept of regeneration ranging from agriculture to architecture has gained importance after the COVID-19 pandemic. Regenerative tourism means going beyond sustainable and responsible tourism. Sustainability is rooted in the principle of meeting present needs without compromising the ability of future generations to satisfy their own needs. The underlying idea of regenerative tourism is, instead, to leave a place in better conditions than it was (Zaman et al., 2023). The scientific debate on regenerative tourism is emerging. Some principles on which regenerative tourism is based include collaborating for virtuous destination management using sustainability standards and new economic performance metrics, defining a maximum number of travelers to avoid over-tourism, promoting the circular use of resources, and protecting the characteristics and identity of the destinations. Examples of sustainable regeneration include the collaboration between local farmers and tourism professionals to restore degraded lands, the integration of tourists with local communities to share traditions, culture, and customs, the introduction of fixed traveler fees to contribute to the costs of natural parks, the incentive use of electric vehicles for trips on the road, and the invitation to visitors to generate social content for promoting the diffusion of local practices on their social media.

Some takeaways that firms can consider for contributing to SDG 12 include:

- defining strategies for travel and tourism businesses based on circular economy models to reduce the production and consumption of services that require high quantities of energy and natural resources;
- developing marketing policies to generate favorable visitor attitudes toward the use of recyclable and biodegradable materials;
- focusing on sustainability, quality, and ethics certificates for a conscious and responsible use of natural resources;
- reconfiguring the business by focusing on regenerative tourism and defining the contents and targets of regeneration;
- introducing comparable and reliable ESG reporting and rating systems based on national and international standards.

In conclusion, the chapter highlights the centrality of tourism in the context of global sustainability objectives and its progressive transformation from an approach based on the mere use of places and services to an orientation toward sustainability to protect places for future generations, up to the innovative vision of regenerative tourism whose task is to actively contribute to the improvement of destinations.

References

Adeola, O., & Evans, O. (2019). Digital tourism: Mobile phones, internet and tourism in Africa. *Tourism Recreation Research, 44*(2), 190–202.

Agyeiwaah, E., McKercher, B., & Suntikul, W. (2017). Identifying core indicators of sustainable tourism: A path forward? *Tourism Management Perspectives, 24*, 26–33.

Ali, A., Rasoolimanesh, S. M., & Cobanoglu, C. (2020). Technology in tourism and hospitality to achieve sustainable development goals (SDGs). *Journal of Hospitality and Tourism Technology, 11*(2), 177–181.

Asmelash, A. G., & Kumar, S. (2019). Assessing progress of tourism sustainability: Developing and validating sustainability indicators. *Tourism Management, 71*, 67–83.

Bastidas-Manzano, A. B., Sánchez-Fernández, J., & Casado-Aranda, L. A. (2021). The past, present, and future of smart tourism destinations: A bibliometric analysis. *Journal of Hospitality and Tourism Research, 45*(3), 529–552.

Becken, S., & Kaur, J. (2021). Anchoring "tourism value" within a regenerative tourism paradigm—a government perspective. *Journal of Sustainable Tourism, 30*(1), 52–68.

Bellato, L., Frantzeskaki, N., Fiebig, C. B., Pollock, A., Dens, E., & Reed, B. (2022). Transformative roles in tourism: Adopting living systems' thinking for regenerative futures. *Journal of Tourism Futures, 8*(3), 312–329.

Bilgihan, A., & Nejad, M. (2015). Innovation in hospitality and tourism industries. *Journal of hospitality and Tourism Technology, 6*(3).

Bolton, D., & Benn, S. (2010). Key concepts in corporate social responsibility. *Key Concepts in Corporate Social Responsibility,* 1–248.

Bramwell, B., Higham, J., Lane, B., & Miller, G. (2017). Twenty-five years of sustainable tourism and the Journal of Sustainable Tourism: Looking back and moving forward. *Journal of Sustainable Tourism, 25*(1), 1–9.

Brida, J. G., Cortes-Jimenez, I., & Pulina, M. (2016). Has the tourism-led growth hypothesis been validated? A literature review. *Current Issues in Tourism, 19*(5), 394–430.

Bynum, B. B., & Muzaffer, U. (2013). Competitive synergy through practicing triple bottom line sustainability: Evidence from three hospitality case studies. *Tourism and Hospitality Research, 13*(4), 226–238.

Carroll, A. B. (2016). Carroll's pyramid of CSR: taking another look. *International Journal of Corporate Social Responsibility, 1*(1), 1–8.

Cave, J., Dredge, D., van't Hullenaar, C., Koens Waddilove, A., Lebski, S., Mathieu, O., ... Zanet, B. (2022). Regenerative tourism: The challenge of transformational leadership. *Journal of Tourism Futures, 8*(3), 298–311.

Cave, J., & Dredge, D. (2020). Regenerative tourism needs diverse economic practices. *Tourism Geographies, 22*(3), 503–513.

Coles, T., Warren, N., Borden, D. S., & Dinan, C. (2017). Business models among SMTEs: Identifying attitudes to environmental costs and their implications for sustainable tourism. *Journal of Sustainable Tourism, 25*(4), 471–488.

Díaz-García, C., González-Moreno, Á., & Sáez-Martínez, F. J. (2015). Eco-innovation: Insights from a literature review. *Innovation, 17*(1), 6–23.

Della Corte, V., Del Gaudio, G., Sepe, F., & Sciarelli, F. (2019). Sustainable tourism in the open innovation realm: A bibliometric analysis. *Sustainability, 11*(21), 6114.

Dwivedi, A., Agrawal, D., & Madaan, J. (2019). Sustainable manufacturing evaluation model focusing leather industries in India: A TISM approach. *Journal of Science and Technology Policy Management, 10*(2), 319–359.

Ettinger, A., Grabner-Krauter, S., & Terlutter, R. (2018). Online CSR communication in the hotel industry: Evidence from small hotels. *International Journal of Hospitality Management, 68*, 94–104.

Gössling, S., Peeters, P., Hall, C. M., Ceron, J. P., Dubois, G., & Scott, D. (2012). Tourism and water use: Supply, demand, and security. An international review. *Tourism Management, 33*(1), 1–15.

Gabor, M. R., Kardos, M., & Oltean, D. F. (2022). Yield management—a sustainable tool for airline E-commerce: Dynamic comparative analysis of E-ticket prices for Romanian full-service airline vs. Low-cost carriers. *Sustainability, MDPI, 14*(22), 1–19 (November).

González-Morales, O., Álvarez-González, J. A., & Sanfiel-Fumero, M. A. (2016). Governance, corporate social responsibility and cooperation in sustainable tourist destinations: The case of the island of fuerteventura. *Island Studies Journal, 11*(2), 561–584, 2016.

Gretzel, U., & Scarpino-Johns, M. (2018). Destination resilience and smart tourism destinations. *Tourism Review International, 22*(3–4), 263–276.

Hassan, S. S. (2000). Determinants of market competitiveness in an environmentally sustainable tourism industry. *Journal of Travel Research, 38*(3), 239–245.

Hatipoglu, B., Ertuna, B., & Salman, D. (2019). Corporate social responsibility in tourism as a tool for sustainable development: An evaluation from a community perspective. *International Journal of Contemporary Hospitality Management, 31*(6), 2358–2375.

Higham, J., & Miller, G. (2018). Transforming societies and transforming tourism: Sustainable tourism in times of change. *Journal of Sustainable Tourism, 26*(1), 1–8.

James, L. (2012). Sustainable corporate social responsibility-An analysis of 50 Definitions for a period of 2000–2011. *ZENITH International Journal of Multidisciplinary Research, 2*(10), 169–193.

Jones, P., Hillier, D., & Comfort, D. (2016). Sustainability in the hospitality industry. *International Journal of Contemporary Hospitality Management, 9.*

Köhler, J., Geels, F. W., Kern, F., Markard, J., Onsongo, E., Wieczorek, A., ... Wells, P. (2019). An agenda for sustainability transitions research: State of the art and future directions. *Environmental Innovation and Societal Transitions, 31*, 1–32.

Kalia, P., Zia, A., & Kaur, K. (2022). Social influence in online retail: A review and research agenda. *European Management Journal.*

Kasemsap, K. (2017). Strategic innovation management: An integrative framework and causal model of knowledge management, strategic orientation, organizational innovation, and organizational performance. In *Organizational culture and behavior: Concepts, methodologies, tools, and applications* (pp. 86–101). IGI Global.

Khan, S. A. R., Zhang, Y., Kumar, A., Zavadskas, E., & Streimikiene, D. (2020). Measuring the impact of renewable energy, public health expenditure, logistics, and environmental performance on sustainable economic growth. *Sustainable Development, 28*(4), 833–843.

Kucukusta, D., Mak, A., & Chan, X. (2013). Corporate social responsibility practices in four and five-star hotels: Perspectives from Hong Kong visitors. *International Journal of Hospitality Management, 34*, 19–30.

Lean, G. L. (2009). Transformative travel: Inspiring sustainability. In R. Bushell, & P. J. Sheldon (Eds.), *Wellness and tourism: Mind, body, spirit, place* (pp. 191–205). Elmsford, NY: Cognizant Communication.

Lee, T. H. (2013). Influence analysis of community resident support for sustainable tourism development. *Tourism Management, 34*, 37–46.

Lenzen, M., Sun, Y. Y., Faturay, F., Ting, Y. P., Geschke, A., & Malik, A. (2018). The carbon footprint of global tourism. *Nature Climate Change, 8*(6), 522–528.

Levy, S. E., & Park, S.-Y. (2011). An analysis of CSR activities in the lodging industry. *Journal of Hospitality and Tourism Management, 18*, 147–154.

Li, M. W., Teng, H. Y., & Chen, C. Y. (2020). Unlocking the customer engagement-brand loyalty relationship in tourism social media: The roles of brand attachment and customer trust. *Journal of Hospitality and Tourism Management, 44*, 184–192.

Li, Z., Wang, D., Abbas, J., Hassan, S., & Mubeen, R. (2022). Tourists' health risk threats amid COVID-19 era: Role of technology innovation, transformation, and recovery implications for sustainable tourism. *Frontiers in Psychology, 12*, 769175.

Lim, C., & McAleer, M. (2005). Ecologically sustainable tourism management. *Environmental Modelling and Software, 20*(11), 1431–1438.

Longo, M. C., & Faraci, R. (2023). Next-generation museum: A metaverse journey into the culture. *Sinergie Italian Journal of Management, 41*(1), 147–176.

Loorbach, D., & Wijsman, K. (2013). Business transition management: exploring a new role for business in sustainability transitions. *Journal of Cleaner Production, 45*, 20–28.

Lozano-Oyola, M., Javier Blancas, F., González, M., & Caballero, R. (2012). Sustainable tourism indicators as planning tools in cultural destinations. *Ecological Indicators, 18*, 659–667.

Madanaguli, A., Srivastava, S., Ferraris, A., & Dhir, A. (2022). Corporate social responsibility and sustainability in the tourism sector: A systematic literature review and future outlook. *Sustainable Development, 30*(3), 447—461.

Madhavan, H., & Rastogi, R. (2013). Social and psychological factors influencing destination preferences of domestic tourists in India. *Leisure Studies, 32*(2), 207—217.

Mantaguti, F., & Mingotto, E. (2016). Innovative business models within niche tourist markets: Shared identity, authenticity and flexibile networks. The case of three Italian SMEs. *Journal of Tourism Research, 6*, 9—10.

Markard, J., Raven, R., & Truffer, B. (2012). Sustainability transitions: An emerging field of research and its prospects. *Research Policy, 41*(6), 955—967.

McWilliams, A., & Siegel, D. (2001). Corporate social responsibility: A theory of the firm perspective. *Academy of Management Review, 26*(1), 117—127.

Mehraliyev, F., Choi, Y., & Köseoglu, M. A. (2019). Progress on smart tourism research. *Journal of Hospitality and Tourism Technology, 10*(4), 522—538.

Pjerotic, L., Delibasic, M., Jokšienė, I., Griesienė, I., & Georgeta, C. P. (2017). Sustainable tourism development in the rural areas. *Transformations in Business and Economics, 16*, 21—30.

Prayag, G., Dookhony-Ramphul, K., & Maryeven, M. (2010). Hotel development and tourism impacts in Mauritius: Hoteliers' perspectives on sustainable tourism. *Development Southern Africa, 27*(5), 697—712, 2010.

Prud'homme, B., & Raymond, L. (2016). Implementation of sustainable development practices in the hospitality industry: A case study of five Canadian hotels. *International Journal of Contemporary Hospitality Management, 28*(3), 609—639.

Pung, J. M., Gnoth, J., & Del Chiappa, G. (2020). Tourist transformation: Towards a conceptual model. *Annals of Tourism Research, 81*.

Pung, J. M., Khoo, C., Del Chiappa, G., & Lee, C. (2022). Tourist transformation: An empirical analysis of female and male experiences. *Tourism Recreation Research*, 1—15.

Purwanda, Eka, & Achmad, Willya (2022). Environmental concerns in the framework of general sustainable development and tourism sustainability. *Journal of Environmental Management and Tourism, 13*, 1911—1917.

Rahman, M. S. U., Simmons, D., Shone, M. C., & Ratna, N. N. (2021). Social and cultural capitals in tourism resource governance: The essential lenses for community focussed co-management. *Journal of Sustainable Tourism*, 1—21.

Rodrigues, M., & Mendes, L. (2018). Mapping of the literature on social responsibility in the mining industry: A systematic literature review. *Journal of Cleaner Production, 181*, 88—101.

Romagosa, F. (2020). The COVID-19 crisis: Opportunities for sustainable and proximity tourism. *Tourism Geographies, 22*(3), 690—694.

Seguí-Amortegui, L., Clemente-Almendros, J. A., Medina, R., & Grueso Gala, M. (2019). Sustainability and competitiveness in the tourism industry and tourist destinations: A bibliometric study. *Sustainability, 11*(22), 6351.

Serra-Cantallops, A., Pena-Miranda, D. D., Ramon-Cardona, J., & Martorell-Cunill, O. (2018). Progress in research on CSR and the hotel industry (2006-2015). *Cornell Hospitality Quarterly, 59*, 15—38.

Sharma, T., Chen, J., & Liu, W. Y. (2020). Eco-innovation in hospitality research (1998—2018): A systematic review. *International Journal of Contemporary Hospitality Management. Diets. Journal of Agricultural and Environmental Ethics, 27* (CA).

Szromek, A. R., & Herman, K. (2019). A business creation in post-industrial tourism objects: Case of the industrial monuments route. *Sustainability, 11*, 1451.

Tsai, H., Tsang, N. K. F., & Cheng, S. K. Y. (2012). Hotel employees' perceptions on corporate social responsibility: The case of Hong Kong. *International Journal of Hospitality Management, 31*, 1143—1154.

UNWTO World Tourism Organization. (2023). *Statistical framework for measuring the sustainability of tourism (sf-mst) Draft prepared for Global Consultation October 2023.* https://www.unwto.org/tourism-statistics/statistical-framework-for-measuring-the-sustainability-of-tourism.

Uyar, A., Koseoglu, M. A., Kuzey, C., & Karaman, A. S. (2023). Does firm strategy influence corporate social responsibility and firm performance? Evidence from the tourism industry. *Tourism Economics, 29*(5), 1272–1301.

Uysal, M., Sirgy, M. J., Woo, E., & Kim, H. L. (2016). Quality of life (QOL) and well-being research in tourism. *Tourism Management, 53,* 244–261.

Viglia, G., & Abrate, G. (2020). Revenue and yield management: A perspective article. *Tourism Review, 75*(1), 294–298.

Waseema, M. (2017). Enhancing destination competitiveness for a sustainable tourism industry: The case of Maldives. *OIDA International Journal of Sustainable Development, 10*(02), 11–24.

World Travel & Tourism Council. (2017). *Environmental, social, and governance reporting in travel and tourism: 3. Sustainability reporting in travel and tourism, ESGs - sustainability reporting in travel and tourism - 2017.pdf.*

Zaman, U., Aktan, M., Agrusa, J., & Khwaja, M. G. (2023). Linking regenerative travel and residents' support for tourism development in Kaua'i Island (Hawaii): Moderating-mediating effects of travel-shaming and foreign tourist attractiveness. *Journal of Travel Research, 62*(4), 782–801.

Website

European Parliament and of the Council. (2022). Corporate sustainability reporting Directive CSRD - (EU) 2022/2464. https://eur-lex.europa.eu/legal-content/EN/TXT/?uri=CELEX%3A32022L2464. (Accessed 31 October 2023).

Europa. (2023). https://ec.europa.eu/commission/presscorner/detail/en/ip_23_3942.

Commission of the European Communities. (2006). *Communication from the commission to the European Parliament, the council and the European economic and social committee implementing the partnership for growth and jobs: Making Europe a pole of excellence on corporate social responsibility, COM(2006) 136 final.* https://eur-lex.europa.eu/LexUriServ/LexUriServ.do?uri=COM:2006:0136:FIN:en:PDF.

Directorate-Directorate-general for internal market, industry, entrepreneurship and SMEs, communication from the commission - towards a common European tourism data space: Boosting data sharing and innovation across the tourism ecosystem.

Global. (2023). *Global destination sustainability index.* https://www.gds.earth/index/.

Meaningful Tourism Index. (2023). *Editor and project group leader: Wolfgang Georg arlt production and cover: Sofija Sarafejeva production assistance: Ana-Maria Tapalaga.* Louise Arditti Published by Meaningful Tourism Center. https://meaningful-tourism.com/wp-content/uploads/2023/05/MEANINGFUL-TOURISM-INDEX-2023.pdf.

One UN for One Planet. (2023). *Input for review sustainable development goal 12.* www.oneplanetnetwork.org).18409Inputs_to_the_review_of_SDG_12_One_UN_for_One_Planet.pdf.

United Nations. (2010). International Recommendations for Tourism Statistics 2008, p. 42. https://unstats.un.org/unsd/publication/seriesm/seriesm_83rev1e.pdf. (Accessed 31 October 2023).

United Nations ten-year framework on sustainable consumption and production patterns.(2023). https://www.unep.org/explore-topics/resource-efficiency/what-we-do/one-planet-network/10yfp-10-year-framework-programmes.

CHAPTER SEVEN

Building sustainable cities through SDG public—private partnerships

1. SDG 11 for sustainable cities and local communities

SDG 11 aims to make cities and communities sustainable to build a more socially, economically, and environmentally equitable future. The evolution toward smart cities represents a necessary step for a safe, resilient, and inclusive city. ICT is a pivotal tool for creating innovative, sustainable, and competitive urban landscapes in terms of quality of life, urban operations management and services, and the preservation of cultural heritage for current and future generations (UNECE—United Nations Economic Commission for Europe, 2020). To monitor progress toward the Sustainable Development Goals (SDGs), the International Telecommunication Union (ITU) in collaboration with other United Nations organizations has developed 92 key performance indicators (core and advanced KPIs) for smart and sustainable cities related to the three dimensions of sustainable development. Sustainability represents a complex challenge, considering the centrality of cities in promoting economic-social development and innovation, in providing opportunities for education, employment, cultural enjoyment and entertainment, as well as in concentrating people and wealth. Various factors influence the planning and reconfiguration of urban spaces in a sustainable way, including rapid urban growth, presence of ethnic groups, mobility needs, services usability, accessibility to public transport systems, and the environmental impact of waste and high carbon dioxide emissions. Governments, institutions, and firms are committed to designing sustainable cities by creating environments favorable to the development of business activities, new jobs, safe and energy-efficient buildings, and resilient economies. Governments at various levels are committed to planning investments in sustainable public works, green public transport and spaces, and participatory and inclusive urban management (Habitat III, 2016). Firms, in turn, can contribute constructively by developing innovative

Being a Sustainable Firm
ISBN: 978-0-443-14062-4
https://doi.org/10.1016/B978-0-443-14062-4.00010-6

solutions and productive processes in line with a responsible and sustainable approach, thus contributing to the overall optimization of urban systems through advanced infrastructure and smart technologies (Longo & Giaccone, 2017). The UN has defined 10 targets and 15 indicators as metrics to monitor the advancements. The SDG 11 targets are: safe and affordable housing, affordable and sustainable transport systems, inclusive and sustainable urbanization, protect the world's cultural and natural heritage, reduction of the adverse effects of natural disasters and the environmental impact of cities, provision of access to safe and inclusive green and public spaces, strong national and regional development planning, implemention of policies for inclusion, resource efficiency, and disaster risk reduction, and support for least developed countries in sustainable and resilient building (The Global Goal, 2023).

Literature (De Jong et al., 2015; Thornbush & Golubchikov, 2020; Yigitcanlar et al., 2019) highlights that the concepts of "sustainable city," "smart city," "ecological city," "low-carbon city," "resilient city," and "knowledge city" have specific and noninterchangeable meanings. In practice, various terms including smart city, green city, or eco–city are used as synonyms for a sustainable city, although they are distinct concepts. Specifically, a smart city refers to the integration of processes, people, and technologies across various urban life aspects, such as transport, healthcare, energy, real estate, utilities, and education to optimize the use of resources, reduce waste, and enhance residents' quality of life. The concept of a smart city is applied to local entities, including small bodies, city, and region, utilizing information technologies to support sustainable economic development (Kulkarni & Farnham, 2016). The distinctive elements of a smart city are technology, innovation, and sustainability aimed at creating the conditions for "smart living," that is, a city characterized by high standards in public health, safety, education, culture, and urban mobility. The use of smart technologies makes the city infrastructure and public services more interconnected, intelligent, and efficient (EASME & DG GROW, 2019). The smart city is based on an ecosystem of technology companies, start-ups, and research institutes and represents an urban development model capable of attracting investments and new business opportunities. A relevant aspect of the smart city is the use of technologies (sensors, digital twins, machine learning, deep learning, smart computing) and digital tools for collecting and analyzing a huge amount of data (Giaccone & Longo, 2016). The use of real–time data enables to optimize logistical flows, improve building energy efficiency, and enhance public safety in the event of calamities and environmental disasters.

A green city is an urban area that aims to guarantee high environmental quality through social responsibility policies and initiatives. Key dimensions are represented by ecological infrastructures, urban ecosystems development, environmental redevelopment, creation of accessible and quality environments, reduction of ecological footprints, mitigation and management of climate change, and sustainable consumption. Strategies to promote green cities emphasize the environmental dimension within the sustainable development goals. They pay particular attention to the quality of environmental resources (air, water, territory/soil and biodiversity), waste reduction, and CO_2 emissions reduction in relation to the urban context of reference (geographical and climatic characteristics, socioeconomic composition of the population), also focusing on citizen involvement.

A sustainable city is designed with the primary objective of contributing to the built environment and related infrastructure, operational functioning, planning and provision of ecosystems and human services, and continuously optimizing efficiency gains (Bibri, 2018). It reconciles respect and protection of the environment with economic, technological, and social progress without compromising future generations (Javidroozi et al., 2023). The fundamental dimensions include sustainable education, renewable energy, waste management, energy efficiency, sustainable transport and logistics, and eco-sustainable construction. These elements, combined with responsible and informed citizens, create the value of sustainability given by quality life improvement and environmental protection, equity, inclusiveness, and social well-being. The Sustainable Cities Index 2022 evaluates, overall, the sustainability performance of 100 global cities chosen from the European capitals considered the most sustainable in the world. Published by Arcadis, the Sustainable Cities Index 2022 proposes an assessment of companies' actions and challenges related to climate change, urbanization, and resources reduction. The 2022 ranking was developed based on 100 global cities with respect to the degree of implementation of the aforementioned challenges that cities face. This ranking presents the cities of Oslo, Stockholm, Tokyo, Copenhagen, Berlin, London, Seattle, Paris, San Francisco, and Amsterdam in the top 10 places. The index considers three essential elements of sustainability (planet, people, and profit) for the classification of cities and uses 51 parameters and 26 indicators to measure these elements. The analysis highlights cities strengths and weaknesses, evaluating urban environment, levels of economic development, and expectations for future growth. The index focuses on *prosperity beyond profit*, linking economic productivity with sustainability pursuit. The indicators were evaluated by experts based on the information available in all cities and on the reliability of the sources. Despite the positive results obtained, the top 10 cities presented excellent values in one of the three areas but none of them reported the highest values in all three indicators. The report differentiates sustainability in one indicator from overall sustainability. This requires a holistic, long-term vision to address the climate emergency, rising inflation and costs of living, environmental disasters, and work-life balance.

2. City systems integration and livability

The sustainable smart city is an urban environment designed to improve the livability of citizens through advanced digital solutions and high-efficiency energy technologies. Through intelligent resources management and systems integration, it outlines a territorial space characterized by digital innovations and information and communication technologies for better environmental quality, social equity, and well-being over the long run. It is characterized by an interactive and responsive administration aiming to meet the needs of interconnectivity infrastructures and human service provisioning coming from the local community (Bibri, 2018; Zimmer et al., 2016). Real-time data communication establishes a continuous link between citizens and administrations to report problems, collect data, or evaluate specific issues.

The smartest city in the world, according to IMD's Smart City Index 2019, is Singapore. The Smart Nation project was implemented in this country to maximize the quality of citizens' lives through the adoption of cutting-edge technological solutions. Based on the 2020 City Motion Index (CIMI), London is in first place among the smart cities in the world, with a leadership in electric mobility solutions. Furthermore, the city hosts the Chief Digital Officer who has the task of guiding the digitalization of the entire city and many innovative technological startups.

Green logistics plays a crucial role in sustainable cities, contributing to environmental sustainability. One of the SDG 11 challenges concerns the use of green transport modes, where emissions quantity is the discriminating variable of the transport means. Due to the global CO_2 impact generated by traditional logistics (Herold & Lee, 2017), green logistics contributes to reduce the negative effects of pollution on the environment, improving the energy efficiency of transport means in relation to the speed, size, and type of goods. Routes redesign and optimization follow the evolution of markets, distribution channels, and technology through a revision of the goods flow, supply location, and transport means (Dekker et al., 2012; Martel & Klibi, 2016; Oberhofer & Dieplinger, 2014). In the context of green logistics, there is also reverse logistics, which deals with the return of material flows to producers through recycling, reuse, and repair with the aim of achieving a balance between economic, environmental, and social objectives. The Logistics Performance Index (LPI) is an efficiency indicator developed by the World Bank to evaluate and compare the distribution logistics performance of different countries with the aim of supporting green supply chain management policies and protecting the environment.

Decision-making in sustainable cities involves variuos subjects. Public institutions and government bodies define territorial planning policies and local development actions. The improvement in environmental performance leads in the short term to an increase in logistics costs due to huge investments for the reconversion of supply activities, redesign of routes, and renewal of transport means. In the long term, efficiency improvement gains can act as an investment catalyst. Some authors (An et al., 2021; Khan et al., 2017) demonstrated that green logistics and green supply chains attract foreign direct investment and foster the development of the renewable energy sector. On the contrary, the lack of green logistics practices may limit the presence of foreign investors (Wanzala & Zhihong, 2016). In the Chinese market, it has been found that green logistics improves firms' economic results in the medium and long term (Zhu & Sarkis, 2004). Finally, some authors (Schniederjans & Hales, 2016) highlighted a positive relationship between eco-sustainable logistics and coordination and integration within the supply chain actors, leading to improvements in financial, operational, and environmental performance.

Firms contribute to sustainable cities through decisions attributable to the strategic, operational, and functional spheres. Strategic decisions concern the number and location of factories, the sizing and saturation of plants and transport means, products supply, and distribution channels. Operational decisions involve planning production and distribution activities and managing goods flows. Functional decisions pertain to logistics resources management, in terms of incoming and outgoing routes, transport methods, and vehicle loading. At firm level, the development of eco-sustainable logistics activities is based on coordination between strategic planning and the management phase to choose the most effective and efficient sustainable logistics routes and flows (Abbasi & Nilsson, 2016; Fahimnia et al., 2015). Collaboration between various public and private actors in planning, designing, and experimenting with green distribution logistics routes and flows is achieved through information sharing, goods flow management systems, plans to reduce vehicles circulation, common storage areas, optimization of transport loading capacity, reduced delivery times, and forecasting of logistics providers.

3. The SDG public—private partnerships

The 2030 Agenda is based on five key concepts: people, planet, prosperity, peace, and partnership. These concepts define the macro areas of action such as the elimination of hunger and poverty to guarantee dignity and equality (people); the respect for nature to protect the environment from the impacts generated by technology and buildings (prosperity); the commitment to promote peaceful, just and inclusive societies to guarantee lasting peace and protect the dignity of peoples (peace); the implementation of stable and solid collaborations and cooperation by combining national and supranational interests to support economic growth (partnership).

Multistakeholder partnership for the SDGs is "an ongoing collaborative relationship among organisations from different stakeholder types aligning their interests around a common vision, combining their complementary resources and competencies and sharing risk, to maximize value creation toward the Sustainable Development Goals and deliver benefit to each of the partners" (SDG Partnership Guidebook 1.11). Starting from this definition, the 2030 Agenda recognizes the importance of interconnection among prosperous businesses, thriving societies, and healthy environments. This involves the development of cooperation and collaboration among citizens, companies, government, NGOs, foundations, the academic world, and institutions. Partnerships involve actors at national, subnational, and city levels to collaborate systematically across different social sectors and achieve a shared vision of the SDGs (Friends of Europe, 2018). The challenge that partnerships face is to encourage transformational development, that is the transformation from unsustainable situations on the economic, social, and environmental levels to situations of sustainability. Unlike traditional local development more oriented toward satisfying the needs and expectations of recipients, transformational development considers recipients as protagonists and critical transformation partners.

The transformation of an inhabited center into a truly sustainable city is a complex and articulated process that requires coordination across various institutional and operational levels, along with adequate funds and financial resources. Small and economically disadvantaged urban centers may struggle to make this type of change. Elements that make an urban context suitable for becoming a sustainable smart city include the production of biogas from efficient waste management; the creation of smart public car parks and hub locations for sustainable mobility; the planning of online parking using booking systems to avoid unnecessary queues and reduce pollution; the management of green and recreational areas according to sustainable criteria; the construction of buildings with low environmental impact consistently with energy efficiency standards; the use of renewable energy and the fossil fuels reduction; the investment in digital telecommunications to monitor and know in advance the flows of city traffic to reduce polluting emissions and facilitate mobility; the diffusion of smart home automation to improve energy efficiency; the adoption of systems for monitoring water consumption and shared real-time information on air quality to allow timely measures to be taken; the use of sensors to collect and analyze the supply chain and product distribution network; the incentives for car sharing systems to reduce environmental impacts; the creation of digital platforms to offer and inform about global, local, and regional ecological services.

Multistakeholder partnerships (MSP) can include a variable number of organizations and define a multitude of collaboration agreements with different methods, contents, and procedures. However, some relevant types of actors can be identified, which constitute partnerships, and which are important in terms of objectives, results, and value generation (United Nations, 2020). They are the following.

Government (Host country). Governments and institutions ensure the democratic representation of people and develop actions and policies to provide national defense, ensure law compliance, and regulate economic activities and taxation. They also play a crucial role in promoting public order, in supporting respect for the environment, in the provision of efficient and effective services, and in the design of sustainable public works. Governments and public bodies collaborate with international organizations and engage investments with private sector support.

Civil society. Civil society includes multiple levels, and it is divided into various organizations with specific purposes, such as social institutions, religious bodies, non-governmental organizations, and citizens. This area, therefore, includes organizations representing different categories such as elderly, women, young people, disabled people, professionals, trade unions, and social centers. Civil society, both as a beneficiary and as a partner, is the subject of various programs based on social equity, inclusion, rights promotion, social and environmental development, support with special services, extension of rights, and democratic representation to the weakest and most disadvantaged people.

Business. Stakeholder business includes the production of goods and services to satisfy public needs or demands. In addition to the profit aim, firms, ranging from micro-businesses to multinationals, carry out actions that have a significant impact on employees, customers, the community, and the environment. They combine resources and knowledge and, through technical innovation, investments and solutions are sources of value creation, contributing significantly to sustainable development.

UN. The United Nations plays a central role in supporting governments and institutions in building and strengthening countries' tools and capacities to implement the national development agenda. The UN pursues several important goals, such as the promotion of sustainable development, the organization of humanitarian aid, respect for international law, the protection of human rights, and support for the maintenance of peace and security.

Foundation. Foundations are divided into different types including corporate foundations, philanthropic foundations, or private foundations. Furthermore, they can be international, national, or subnational, depending on their areas, interests, goals, and resources. Their activity includes financial support or program management assistance through their own funding sources.

Other stakeholders. Other important categories in creating partnerships are universities, social media, influencers, tribes, trade unions, members of institutions and local authorities, politicians, and administrators. Fig. 7.1 shows the stakeholder resources map.

Partnerships can create value when their action is consistent with the set objectives and generate benefits and value for each individual partner. The process of building a partnership is not linear or regular; instead, it is very complex and requires strong coordination to achieve a precise path of training, implementation, and alignment among partners. The partnership formation process identifies actions to be taken to align the parties and stipulate an agreement to make the partnership operational. The phases can be summarized in the following activities: alignment of interests; definition of an overall vision and purpose/mission; planning of objectives and activities; identification of roles and responsibilities; indication of the partnership basic requirements; signature of the partnership agreement; management of the partnership; and contribution of the various parties involved. Complex relationships develop within the partnership, creating continuous and changing interactions between the subjects. The institutional relations system is

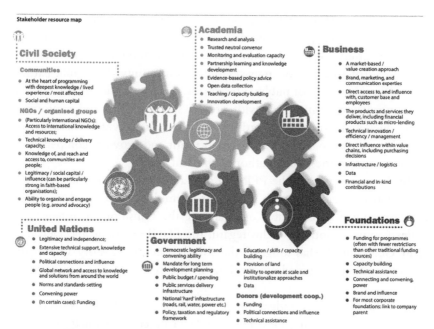

Figure 7.1 The stakeholders map. *From United Nations (2022).*

based on trust, transparency, balanced power, mutual and fair benefit, commitment, and responsibility.

In the context of partnership, the public sector's ability to create relationships with interlocutors is important. In fact, through political cycles and the administration of public spending, governments and public institutions develop valuable relationships with business organizations, often facilitated by forums of public—private dialogue and intermediation activities. Private—public partnerships, which involve sectors and stakeholders in economic-financial, technological, and commercial areas, are expanding to implement the 2030 Agenda plan. Cooperation between the public and private sectors takes shape through different forms of medium-long term SDG public—private partnerships (SDGPPP), which can be realized through interventions and contracts.

The PPP is a public—private collaboration agreement for the design, offer, and financing of public services or works managed in accordance with contractual provisions by private companies selected through a tender (Marsilio et al., 2011; Hodge, 2007; Linder, 1999). PPPs typically concern complex and innovative public works projects or the provision of public utility services, with the private sector involvement in different project phases and in fundraising. The projects include economic infrastructures and public utility service networks, social, cultural, and recreational facilities, green infrastructure, and open spaces for the community. The public partner defines the objectives of public interest, the quality of the services offered, the control over compliance with project specifications, and conformity to product and process standards. The service is provided for a period enough to amortize all costs and make a profit and provides compensation to the firm from the public authority and end users.

In the coming years, the PPP will play a key role in supporting public investments, especially in the sharing of technologies, financial resources, and professional skills capable of catalyzing the transition to sustainability in developing countries. In this regard, the SDGF—Sustainable Development Goals Fund—has assigned the Private Sector Advisory Group the task of assisting public administrations in the selection and activation of PPPs.

In light of these considerations, the role of governments is becoming increasingly strategic in achieving the SDG 11 objectives by activating policies and incentives that steer the different categories of partners towards transformational development based on accountability and commitment. Sustainability, in this context, is related to the concept of accountability, which considers the ability and duty to provide an account for the actions for which an organization is responsible in social, economic, and environmental terms. The role of the public sector in the pursuit of sustainability is based on the concept of public accountability, understood as responsibility toward the social community. Central governments represent the public interest while public organizations, including universities, local authorities, and government institutions, plan, decide, and act in line with this interest with the aim of achieving social and environmental objectives and institutional policies.

The goals and boundaries of public accountability have evolved in relation to social, cultural, and political institutional changes, expanding its meaning and investigation perspectives (Mulgan, 2000; Rosair & Taylor, 2000). Furthermore, this evolution also has interested the adoption and implementation of new communication and evaluation tools systems. Sustainable reporting activities provide accessible information regarding the acquisition and use of resources (allocative efficiency), the level of services produced (intermediate results or outputs), and the final impacts of these services on local realities (final results). Additionally, it explains and clarifies the causal relationships between resources used, outputs, and outcomes in social and environmental terms. Measuring and reporting results of administrative activity are declined with respect to the recipients of the information and represent a tool for social planning and control, internal and external governance, and public communication. The public sector plays an essential role in identifying new ways of social and environmental development. In this sense, some authors (Elkington, 2006; Bolton, 2021) stated that the role of governments and public administrations in planning and regulating the sustainable activities carried out by the economic world concerns various areas such as invasion, internalization, inclusion, integration, and incubation. This is because, with the emergence of sustainable technologies and practices, firms "invade" transversally different fields with multiple economic, environmental, and social impacts, attributable to their "invasion." The government also takes an important role in the internalization process, attempting to ensure that positive externalities are adequately valued and internalized. Even in the process of "inclusion" of stakeholders, the public sector is committed to establishing actions and collaborations with interlocutors directly and indirectly involved in decision-making for sustainable city development. In the "integration" of the triple bottom line elements, governments play a notable role as well as in the "incubation" processes by encouraging the establishment of start-ups and new ventures considered more sustainable and by creating favorable conditions for accelerated growth.

4. SDG evaluation methodology in public—private partnerships

The sustainability pursuit requires the implementation of different forms of partnership, which has stimulated the adoption of tools and methods to evaluate the sustainability of infrastructure projects and initiatives for the local community implemented by PPPs. The Economic and Social Council (Economic Commission for Europe, Committee on Innovation, Competitiveness and Public-Private Partnerships, 2023) has introduced a methodology to evaluate the sustainable development objectives of PPP, defining the PPP and Infrastructure Evaluation and Rating System (PIERS). The Guiding Principles define PPPs for the SDGs according to specific areas such as access and equity; economic effectiveness and fiscal sustainability; environmental sustainability and resilience; replicability; and stakeholder engagement. The PIERS methodology is applicable to all PPP projects

and models across different sectors and geographical locations. It is based on three evaluation elements:
— definition of criteria and indicators to measure the achievement of sustainable development objectives in the five areas mentioned above;
— outcomes weighting system;
— attribution of a score to the PPP partners giving insights regarding project's revisions to make them more compliant with their objectives.

The PIERS rating system has the characteristics of flexibility and adaptability to all types and sizes of PPPs. It is applicable to different territorial contexts and countries. The system is measurable as it includes qualitative and quantitative methods to measure the SDG results achieved by PPPs in different areas. The methodology uses a broad set of 22 criteria and 68 indicators to evaluate the incremental effects in line with the SDGs and a shared language between the public and private sectors to have a coherent assessment of the PPPs' contribution to SDG results. Specifically, the evaluation criteria of the PIERS' areas are the following.

Access and equity. This area aims to improve access to essential public services and reduce poverty to support social development and limit negative impacts on livelihoods and social well-being. Equity is ensured by providing equal access to PPP services through proactive measures for disadvantaged subjects. The criteria identified to evaluate performance are as follows:
— essential services provided;
— degree of affordability and universal access;
— growth in the level of equity and social justice;
— long-term access and equity plan;
— mitigation and elimination of physical and economic displacement.

Economic effectiveness and fiscal sustainability. This area concerns the project's contribution to economic growth and quality employment, as well as an economic evaluation of the PPP option compared to other contractual forms of public service procurement. It evaluates the project's ability to efficiently use economic resources and generate a reasonable level of profitability from affordable tariffs and sustainable management of the PPP budget and public debt. The criteria for evaluating project performance against the economic effectiveness and fiscal sustainability outcomes are as follows:
— degree of corruption and practices to encourage transparent procurement;
— maximizing economic feasibility and fiscal sustainability;
— maximizing long-term financial viability;

— increase of the level of employment and economic opportunities.

Environmental sustainability and resilience. This area concerns the protection of the planet, the preservation of biodiversity, and the challenge of climate change. Resilience in this context refers to the ability of a community to resist risks and adapt and transform in a timely and effective manner. The criteria for evaluating the performance of the project with respect to this area are as follows:

— reduction of greenhouse gas emissions and improvement of energy efficiency;
— waste reduction and redevelopment of areas and land;
— limitation of water consumption and wastewater discharge;
— protection of biodiversity;
— risk assessment and management of environmental disasters.

Replicability. This area concerns the ability of a project to be replicated, either in whole or in part, in line with international best practices and sustainable development goals. The criteria relating to this area are as follows:

— degree of replicability and scalability;
— standardization of models and procedures for PPP tenders;
— strengthening government, industry, and community capacity;
— support for innovation and technology transfer.

Stakeholder engagement. This area refers to the public participation of all parties potentially interested in the project, through their effective and inclusive involvement in the PPP decision-making processes. The criteria for evaluating this area are as follows:

— stakeholder engagement and public participation plan;
— maximizing public participation;
— transparency and quality of information on the project;
— management of public complaints and end users' feedback.

Weighting and scoring. The criteria and indicators were developed to assign a score to the five PPPs and evaluate their performance against the SDGs, taking into account their correlation with the three pillars of sustainable development (economic, social, and environmental pillars).

Implementation and self-evaluation. The scores obtained through the PIERS methodology represent the basis for reviewing and improving PPP projects.

Public sector sustainability reporting standards enhance the achievement of the PIERS' areas concerning transparency, the government's social responsibility for the long-term effects of public interventions, and the broad and informed participation of stakeholders in the PPPs' decision-making process. In 2022, the IPSASB—International Public Sector Accounting

Standards Board—has initiated public sector specific sustainability reporting projects to be adopted as guidelines on the following priority topics: general requirements for disclosure of sustainability-related financial information, climate-related disclosures, and nonfinancial disclosures for natural resources. Furthermore, from 2023, the IPSASB is committed to developing a specific climate disclosure standard for the public sector. This document outlines the sustainability reporting requirements applicable to all central government bodies falling under the Greening Government Commitments (GGC) and reporting in accordance with HM Treasury's Government Financial Reporting Manual (FReM). Sustainability reporting ensures transparency of public sector sustainability performance and compliance with reporting requirements.

5. Takeaways for sustainable PPP managers: Thinking about nature-based solutions and natural step approach

SDG 11 is a complex objective to achieve because it requires interaction between public institutions, businesses, and communities to make cities sustainable and to create an inclusive, fair, and cohesive urban context. The evolution from smart cities to sustainable cities represents a fundamental turning point in building cities of the future where citizens can live in a safer, healthier, and more technological environment. Achieving SDG 11 represents a challenge and an active commitment by stakeholders operating in various sectors at all levels, central institutions, local governments, and citizens (Cardillo & Longo, 2020). A widespread tool in encouraging the participation of public and private actors is the sustainable PPP. Although there is a consensus on different forms and models of PPP, a relevant issue to which decision-makers are called is to identify shared approaches and common solutions to promote sustainability in a holistic vision and not just for individual pillars (planet, people, profit). However, the different functional aspects of sustainable management have led in some cases to the failure of sustainability practices (Ben Aissa & Goaied, 2017). The Nature-based Solution (NBS) and the Natural Step (NS) represent two innovative approaches that can guide managers and public actors in outlining and implementing solutions based on nature.

The NBS is a promising alternative theoretical approach where nature is central to evaluate and implement processes, providing simultaneous environmental, social, and economic benefits (Baldarelli & Cardillo, 2022;

Ma et al., 2022). Decisions revolve around nature, and therefore the way in which human actions and choices interact with nature represents the basis for designing sustainable cities, considering it as a productive resource of social benefits (Chen et al., 2023; Liquete et al., 2016; Mabrouk et al., 2023; Olya & Alipour, 2015).

Favored by cooperation among public bodies, businesses, and citizens, NBS concerns interventions that push toward the diffusion of greenery in cities and extraurban spaces, the protection of biodiversity, and the resilience and livability of the urban context (Castellar et al., 2022; Li et al., 2022). NBS brings elements and characteristics of nature to landscapes and cities through economical, adaptable, multipurpose, and durable engineering solutions. Urban planning interventions concern, for example, rain gardens, green walls, tactile museum gardens, and scattered vegetation in the urban and extraurban centers based on low-carbon technological innovation (Raymond, Berry, et al., 2017; Raymond, Frantzeskaki, et al., 2017). Fig. 7.2 shows the system of indicators, methods, and challenge areas for assessing the impacts of NBS.

A circular and flexible scheme emerges from Fig. 7.1, which underlines the importance of a holistic approach to the design, implementation, and evaluation of NBS, which is divided into three main phases: understanding the environmental and socioecological context of the design; designing NBS based on the interconnectedness of multiple challenges; implementing NBS at multiple scales using adaptive management in response to emerging risks; monitoring and evaluating NBS through multiactor coproduction processes capitalizing on cobenefits. An important phase of this process is the participation of various stakeholders to evaluate alternative paths and promptly collect feedback.

The natural step (NS) is an international nonprofit organization that supports strategic decision-making for achieving sustainable development (Hämäläinen et al., 2023; Hembd & Silberstein, 2010; James & Lahti, 2004; Rafei et al., 2022; Upham, 2000). The NS methodology is based on a shared and common language to manage the change toward sustainability in a simple and effective way. The process realized through this approach has specific characteristics, that are: Science-based (compliant with the relevant scientific peer-reviewed journals); Necessary (considering necessary conditions for a sustainable society); Comprehensive (concerning the whole system and the cause–effect relationship); Universal (appropriate to all fields and expertise); Concrete (applying to day-to-day problem-solving); Distinct (nonoverlapping parameters).

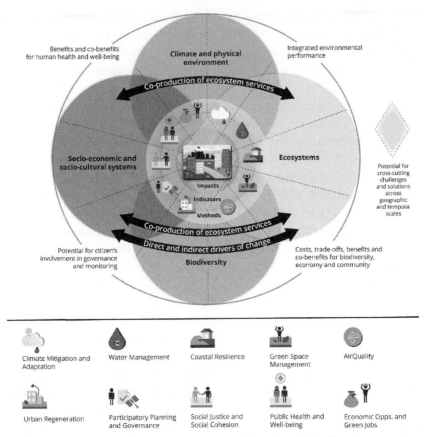

Figure 7.2 The nature-based solution (NBS) indicators and challenge areas. *From Raymond, Frantzeskaki, et al. (2017).*

The NS implementation path follows the sequence ABCD— Awareness, Baseline assessment, Creative solutions, Devise a plan (Fig. 7.3). Awareness refers to the knowledge of the meaning of sustainability, planet, and stakeholders through a clear, simple, and systemic vision. The Baseline Assessment performs a sustainability gap analysis of the organization's key flows and impacts, identifying critical issues and business implications. Creative Solutions lead to the creation of sustainable products, services, organizations, and partnerships. The Devise a plan step involves defining priorities to design a flexible and advantageous path toward sustainability.

Sustainable PPP managers are called upon to reflect on how to implement nature-based approaches. Some useful takeaways to consider include:

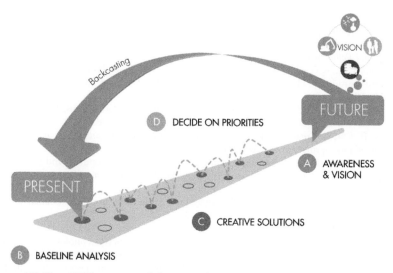

Figure 7.3 The ABCD process of the natural step approach. *From The Natural Step, www.thenaturalstep.org/approach, 2023.*

— knowing in depth the state of each territory to have a picture of the areas in which interventions can be carried out in harmony with the place characteristics;

— aligning existing urban planning strategies and governance processes with NBS actions;

— adopting strategies that systematically address challanges such as climate change, urban regeneration, social cohesion and inclusion, environmental management, food safety, water shortage, ecosystem services; green infrastructure; natural disaster planning;

— selecting appropriate partners to develop innovative and implementable NBS and defining tools for stakeholder involvement and participation;

— managing transition processes by designing NBS in line with the ecosystems of different sectors.

In conclusion, the chapter outlines strategies and solutions that firms can develop in partnership with public institutions to contribute in a responsible way to making their business, territories, and urban contexts more sustainable.

References

Abbasi, M., & Nilsson, F. (2016). Developing environmentally sustainable logistics: Exploring themes and challenges from a logistics service providers' perspective. *Transportation Research Part D: Transport and Environment, 46,* 273–283.

An, H., Razzaq, A., Nawaz, A., Noman, S. M., & Khan, S. A. R. (2021). Nexus between green logistic operations and triple bottom line: Evidence from infrastructure-led Chinese outward foreign direct investment in belt and road host countries. *Environmental Science and Pollution Research, 28*(37), 51022−51045.

Baldarelli, M. G., & Cardillo, E. (2022). Managerial paths, social inclusion, and NBS in tactile cultural products: Theory and practice. *Journal of Hospitality and Tourism Research, 46*(3), 544−582.

Ben Aissa, S., & Goaied, M. (2017). Performance of tourism destinations: Evidence from Tunisia. *Journal of Hospitality and Tourism Research, 41*(7), 797−822.

Bibri, S. E. (2018). Smart sustainable cities of the future. In *The urban book series*. Cham, Switzerland: Springer International Publishing.

Bolton, M. (2021). Public sector understanding of sustainable development and the sustainable development goals: A case study of Victoria, Australia. *Current Research in Environmental Sustainability, 3*, 100056.

Cardillo, E., & Longo, M. C. (2020). Managerial reporting tools for social sustainability: Insights from a local government experience. *Sustainability, 12*(9), 3675.

Castellar, J. A., Torrens, A., Buttiglieri, G., Monclus, H., Arias, C. A., Carvalho, P. N., ... Comas, J. (2022). Nature-based solutions coupled with advanced technologies: An opportunity for decentralized water reuse in cities. *Journal of Cleaner Production, 340*, 130660.

Chen, H. S., Lin, Y. C., & Chiueh, P. T. (2023). Nexus of ecosystem service-human health-natural resources: The nature-based solutions for urban PM2.5 pollution. *Sustainable Cities and Society, 91*, 104441.

De Jong, M., Joss, S., Schraven, D., Zhan, C., & Weijnen, M. (2015). Sustainable−smart−resilient−low carbon−eco−knowledge cities; making sense of a multitude of concepts promoting sustainable urbanization. *Journal of Cleaner Production, 109*, 25−38.

Dekker, R., Bloemhof, J., & Mallidis, I. (2012). Operations research for green logistics−an overview of aspects, issues, contributions and challenges. *European Journal of Operational Research, 219*(3), 671−679.

EASME, DG GROW. (2019). *Digital cities challenges: Designing digital transformation strategies for EU cities in the 21st century. Final report.* Luxembourg: Publications OCE of the European Union. https://www.intelligentcitieschallenge.eu/sites/default/les/2019-09/EA-04-19-483-EN-N.pdf.

Economic and Social Council, Economic Commission for Europe. (2023). *ECE/CECI/2023/4 committee on innovation, competitiveness and public-private partnerships, public-private partnerships and infrastructure evaluation and rating system (PIERS): An evaluation methodology for the sustainable development goals.* https://unece.org/sites/default/files/2023-04/ECE_CECI_2023_4_2305092E.pdf.

Elkington, J. (2006). Governance for sustainability. *Corporate Governance: An International Review, 14*(6), 522−529.

Fahimnia, B., Bell, M. G. H., Hensher, D. A., & Sarkis, J. (2015). The role of green logistics and transportation in sustainable supply chains. In B. Fahimnia, M. Bell, D. Hensher, & J. Sarkis (Eds.), *Green logistics and transportation. Greening of industry networks studies* (Vol. 4). Cham: Springer.

Friends of Europe. (2018). *Agenda 2030 and public private partnership.* https://www.friendsofeurope.org/wp/wp-content/uploads/2019/04/2018_foe_dpf_pub_agenda-2030_web.pdf.

Giaccone, S. C., & Longo, M. C. (2016). Insights on the innovation hub's design and management. *International Journal of Technology Marketing, 11*(1), 97−119.

Hämäläinen, R. M., Halonen, J. I., Haveri, H., Prass, M., Virtanen, S. M., Salomaa, M. M., ... Haahtela, T. (2023). Nature step to health 2022−2032: Interorganizational collaboration to prevent human disease, nature loss, and climate crisis. *The Journal of Climate Change and Health, 10*, 100194.

Habitat, III. (2016). New urban agenda. Available from: https://unhabitat.org/sites/default/les/2019/05/nua-english.pdf.

Hembd, J., & Silberstein, J. (2010). *Sustainability and community development. Introduction to community development: Theory, practice, and service-learning.* California, USA: Sage Publications.

Herold, D. M., & Lee, K. H. (2017). Carbon management in the logistics and transportation sector: An overview and new research directions. *Carbon Management, 8*(1), 79–97.

Hodge, G. A., & Greve, C. (2007). Public-private partnerships: An international performance review. *Public Administration Review, 67,* 545–558.

James, S., & Lahti, T. (2004). *The natural step for communities: How cities and towns can change to sustainable practices.* New Society Publishers.

Javidroozi, V., Carter, C., Grace, M., & Shah, H. (2023). Smart, sustainable, green cities: A state-of-the-art review. *Sustainability, 15*(6), 5353.

Khan, S. A. R., & Qianli, D. (2017). Does national scale economic and environmental indicators spur logistics performance? Evidence from UK. *Environmental Science and Pollution Research, 24,* 26692–26705.

Kulkarni, P., & Farnham, T. (2016). Smart city wireless connectivity considerations and cost analysis: Lessons learnt from smart water case studies. *IEEE Access, 4,* 660–672.

Linder, S. H. (1999). Coming to terms with the public-private partnership—A grammar of multiple meanings. *American Behavioral Scientist, 43,* 35–51.

Li, H., Peng, J., Jiao, Y., & Ai, S. (2022). Experiencing urban green and blue spaces in urban wetlands as a nature-based solution to promote positive emotions. *Forests, 13*(3), 473.

Liquete, C., Udias, A., Conte, G., Grizzetti, B., & Masi, F. (2016). Integrated valuation of a nature-based solution for water pollution control. Highlighting hidden benefits. *Ecosystem Services, 22,* 392–401.

Longo, M. C., & Giaccone, S. C. (2017). Struggling with agency problems in open innovation ecosystem: Corporate policies in innovation hub. *The TQM Journal, 29*(6), 881–898.

Ma, S., Wang, H. Y., Zhang, X., Wang, L. J., & Jiang, J. (2022). A nature-based solution in forest management to improve ecosystem services and mitigate their trade-offs. *Journal of Cleaner Production, 351,* 131557.

Mabrouk, M., Han, H., Fan, C., Abdrabo, K. I., Shen, G., Saber, M., ... Sumi, T. (2023). Assessing the effectiveness of nature-based solutions-strengthened urban planning mechanisms in forming flood-resilient cities. *Journal of Environmental Management, 344,* 118260.

Marsilio, M., Cappellaro, G., & Cuccurullo, C. (2011). Intellectual structure of PPPs research: A bibliometric analysis. *Public Management Review, 13,* 763–782. ISSN: 1471-9037.

Martel, A., & Klibi, W. (2016). *Designing value-creating supply chain networks.* Cham: Springer.

Mulgan, R. (2000). 'Accountability': An ever-expanding concept? *Public Administration, 78*(3), 555–557.

Oberhofer, P., & Dieplinger, M. (2014). Sustainability in the transport and logistics sector: Lacking environmental measures. *Business Strategy and the Environment, 23*(4), 236–253.

Olya, H., & Alipour, H. (2015). Developing a climate-based recreation management system for a Mediterranean Island. *Fresenius Environmental Bulletin, 24*(12), 1–33.

Rafei, M., Esmaeili, P., & Balsalobre-Lorente, D. (2022). A step towards environmental mitigation: How do economic complexity and natural resources matter? Focusing on different institutional quality level countries. *Resources Policy, 78,* 102848.

Raymond, C. M., Berry, P., Breil, M., Nita, M. R., Kabisch, N., de Bel, M., ... Lovinger, L. (2017a). *An impact evaluation framework to support planning and evaluation of nature-based solutions projects. Report prepared by the EKLIPSE expert working group on nature-based solutions to promote climate resilience in urban areas.* Wallington, United Kingdom: Centre for Ecology and Hydrology.

Raymond, C. M., Frantzeskaki, N., Kabisch, N., Berry, P., Breil, M., Razvan Nita, M., Geneletti, D., & Calfapietra, C. (2017b). A framework for assessing and implementing the co-benefits of nature-based solutions in urban areas. *Environmental Science and Policy, 77*, 15–24.

Rosair, M., & Taylor, D. W. (2000). The effects of participating parties, the public and size on government departments' accountability disclosures in annual reports. *Accounting, Accountability and Performance, 6*(1), 77–97.

Schniederjans, D. G., & Hales, D. N. (2016). Cloud computing and its impact on economic and environmental performance: A transaction cost economics perspective. *Decision Support Systems, 86*, 73–82.

The Global Goal. (2023). https://www.globalgoals.org/goals/11-sustainable-cities-and-communities/.

The Natural Step. https://thenaturalstep.org/approach/5-levels/.

Thornbush, M., & Golubchikov, O. (2020). *Sustainable urbanism in digital transitions, from low carbon to smart sustainable cities.* Cham: Springer.

United Nation. (2020). *The SDG partnership Guidebook: A practical guide to building high impact multi-stakeholder partnerships for the sustainable development goals, Darian Stibbe and Dave Prescott, the partnering initiative and UNDESA 2020* (p. 33). https://sdgs.un.org/sites/default/files/2022-02/SDG%20Partnership%20Guidebook%201.11.pdf Accessed 16 November 2023.

United Nations Economic Commission for Europe. (2020). *UNECE, people-smart sustainable cities.* https://unece.org/sites/default/files/2021-01/SSC%20nexus_web_opt_ENG_0.pdf.

Upham, P. (2000). An assessment of the Natural Step theory of sustainability. *Journal of Cleaner Production, 8*(6), 445–454.

Wanzala, W. G., & Zhihong, J. (2016). Integration of the extended gateway concept in supply chain disruptions management in East Africa-Conceptual paper. *International Journal of Engineering Research in Africa, 20*, 235–247.

Yigitcanlar, T., et al. (2019). Can cities become smart without being sustainable? A systematic review of the literature. *Sustainable Cities and Society, 45*, 348–365.

Zhu, Q., & Sarkis, J. (2004). Relationships between operational practices and performance among early adopters of green supply chain management practices in Chinese manufacturing enterprises. *Journal of Operations Management, 22*(3), 265–289.

Zimmer, K., Fröhling, M., & Schultmann, F. (2016). Sustainable supplier management—a review of models supporting sustainable supplier selection, monitoring and development. *International Journal of Production Research, 54*(5), 1412–1442.

Conclusions: Sustainable Development Goals implications and interconnections on the sustainability management

The growing awareness toward sustainability issues launches ambitious objectives both in the strategies to be followed and in the adoption of reliable performance evaluation systems. The creation of sustainable and lasting value for stakeholders, starting from shareholders, depends on firms' ability of firms to balance the three dimensions of sustainability. This involves engaging corporate areas in Sustainable Development Goals (SDGs), making actors responsible in decision-making processes, and fostering long-term growth.

The book underlines the centrality of sustainability issues in business strategies that start from the corporate level and are reflected in the macroeconomic, social, and environmental context. It also highlights the importance of greater integration and diffusion of SDG public—private partnerships among the main actors of the economic, political, and regulatory context, including governments and institutions, policymakers, standard setters, managers, universities, and research centers, to define and implement strategic sustainability objectives. Additionally, the volume reinforces the relevance of disclosure, reporting, communication, and dissemination of nonfinancial information, providing for a broad involvement of the stakeholders.

Communicating firms' sustainable action plans allows third parties to understand the extent of corporate social responsibility and its contribution to environmental and social impact improvement. Beyond enhancing the brand's green reputation, the commitments undertaken in the context of Corporate Social Responsibility (CSR) are pivotal in exploring new sustainable business models. The results highlight the importance of outlining an integrated thinking and reporting approach that enhances the quality of financial and nonfinancial information. This enables companies and stakeholders to make more informed decisions on the firm's ability to create, destroy, or preserve value in the long term.

The seven chapters covered in the book provide a mapping of the several dimensions of sustainability and the way in which both firms and organizations move within this framework.

The first chapter outlined the boundaries of sustainability by presenting approaches and regulations to promote economic growth and a more resilient, sustainable, and inclusive society. The various circular models applied to industrial production processes open new ways for companies to grow and innovate and define business strategies that include economic, environmental, and social value. The dominant themes addressed in this chapter were sustainability, regulation, circular economy models, and strategies for being or becoming sustainable.

The second chapter analyzed the EU SDG objectives and then focused on SDGs 9, 11, and 12. The analysis highlighted how companies, directly involved in the pursuit of these goals, contribute to the creation of sustainable production processes and respond to community expectations from an inclusive, green, and resilient perspective. Dominant themes addressed in this chapter were the SDGs' content and monitoring, the European Green Deal, responsible industrialization, sustainable cities, sustainable consumption, and production models.

The third chapter focused on SDG indicators for sustainability assessment. The analysis highlighted their trend based on territorial extension, their progress compared to the EU average, and the importance of national and international guidelines to make sustainability measures applicable and implementable on the basis of four areas (economic, social, environmental, and institutional). Dominant themes covered in this chapter were the SDG indicators set, contents and trend, the fair and sustainable wellbeing (BES) indicators, the EU indicators for goal, and the multipurpose indicators.

The fourth chapter examined standards and guidelines leading corporate sustainability reporting from transition to its actual implementation at various corporate levels. The study highlighted how reporting and disclosure of nonfinancial information support stakeholders in investment decisions and financial risk mitigation. Dominant themes were the sustainability reporting standards and guidelines, the nonfinancial information disclosure, the global report initiative (GRI) framework, the environmental, social and governance (ESG) pillars, and corporate sustainability reporting.

Chapter five connected SDG 9 with the fashion sector. The study examined how fashion firms address the strategic issue of sustainability, moving from traditional production models to circular systems. Dominant themes were innovation, sustainability-based strategies, certifications, the circular model to the product life cycle, and Sustainable Eco-Chic Fashion Tech.

Chapter six linked SDG 12 to the tourism sector. In addition to the contribution of digital technologies to support sustainable production and consumption models, the study discussed the main differences between corporate social responsibility (CSR) and environmental and social governance (ESG). Dominant themes were sustainable tourism, digital tourism, CSR, tourism sustainability reporting, and regenerative tourism.

Chapter seven linked SDG 11 to sustainable cities and communities within five key concepts of Agenda 2023 (People, Planet, Prosperity, Peace, and Partnership). The study outlined the sustainable smart city, as an innovative urban context designed to improve the quality of life of citizens, thanks to interconnected environmental, digital, and logistical solutions. The SDG public—private partnership plays a strategic role in achieving shared planning of urban areas. Dominant themes were sustainable and smart cities, the SDG public—private partnerships, the UNECE PPP and infrastructure evaluation and rating system (PIERS) methodology, the Nature-based solution, and the Nature Step.

At the conclusion of this volume, meaningful strategic and managerial takeaways emerge.

- A sustainability-oriented company bases its entire strategy on the concept of regeneration to improve as well as to protect the planet.
- Being born sustainable is different from becoming sustainable. A firm that is born sustainable plans its business by evaluating social, environmental, and economic impacts and activating a dynamic eco-sustainable circuit. A firm already operating on the market poses the strategic question of revising its paths and objectives and introducing competitive levers in line with sustainable development.
- When formulating their sustainability strategies, companies could start by exploring the relevant SDG in terms of *what it is, what it consists of, how to decline, what the role of firms, how to monitor the firms' contribution, and trend to date.*
- Focus on an integrated system of national and international indicators to ensure the coherence and comparability of the information processed and a shared evaluation of company policies, highlighting targets, interrelationships, and multipurpose indicators.
- When choosing sustainability standards, organizations need to take into account their diffusion in the sector, their consistency with the activity carried out, and the extent of investments, priorities, and actions that the company is called upon to carry out.
- Identify criteria that guide firms in defining and representing the sustainable business model in an organic, solid and coherent way with the value creation process. This involves focusing on impact assessment, materiality analysis, and performance analysis and ESG rating.

- Select the material issues on which the sustainability strategy is based and evaluate the credibility of the methodology used in terms of identification, selection and social significance, considering both environmental and economic items.
- Fashion companies can contribute to SDG 9 by applying circular models to the product life cycle and pursuing an orientation toward Sustainable Eco-Chic Fashion Tech.
- Tourism businesses can contribute to SDG 12 by focusing on regenerative tourism, defining the contents and objectives of regeneration and findings ways to improve destinations.
- In achieving SDG 11, public—private partnerships implement urban planning strategies and land governance processes according to the Nature-based solution and Nature Step approaches.

In conclusion, being a sustainable firm means finding the coordinates to scale sustainability dimensions by undertaking virtuous paths in harmony with the planet.

Index

Note: 'Page numbers followed by f indicate figures, t indicates tables'

Printed and bound by CPI Group (UK) Ltd, Croydon, CR0 4YY

08/05/2025

01864775-0002